THE HUMAN INHERITANCE

Genes, Language, and Evolution

Edited by

BRYAN SYKES
Institute of Molecular Medicine,
University of Oxford

OXFORD
UNIVERSITY PRESS

OXFORD
UNIVERSITY PRESS

Great Clarendon Street, Oxford OX2 6DP

Oxford University Press is a department of the University of Oxford
and furthers the University's aim of excellence in research, scholarship,
and education by publishing worldwide in

Oxford New York

Athens Auckland Bangkok Bogotá Buenos Aires Calcutta
Cape Town Chennai Dar es Salaam Delhi Florence Hong Kong Istanbul
Karachi Kuala Lumpur Madrid Melbourne Mexico City Mumbai
Nairobi Paris São Paulo Singapore Taipei Tokyo Toronto Warsaw

and associated companies in Berlin Ibadan

Oxford is a registered trade mark of Oxford University Press

Published in the United States
by Oxford University Press Inc., New York

A catalogue record for this book is available from the British Library

Library of Congress Cataloging in Publication Data
The human inheritance / edited by Bryan [sic] Sykes.
Includes bibliographical references and index.
1. Human genetics. 2. Human evolution. 3. Language and
languages–Origin. 4. Evolutionary genetics. I. Sykes, Brian D.
QH431.H8374   1999   599.93'5–dc21   99–15223

ISBN 0 19 850274 5 (Hbk)

Typeset by Footnote Graphics, Warminster, Wilts

Printed in Great Britain
by
Biddles Ltd, Guildford & King's Lynn

Preface

Very little excites the human curiosity quite so much as contemplating our own origins. More than any other branch of science, evolution, and human evolution in particular, is capable of bringing out the best—and the worst—in its professional practitioners. Working from what is essentially the same data, schools of opinion come to diametrically opposed conclusions. Are we just adapted Neanderthals or a new species altogether which wiped them out? Were our ancestors in Britain hunter-gatherers or were these icons of the collective subconscious swept aside by a flood of farmers? Did the first Americans enter the continent 30 000 or 12 000 years ago? Did the first Polynesians sail against wind and current to an unknown fate or were they just blown across from South America while out fishing? And why do we speak different languages? Is it because language also traces our biological history, or are the two things completely unrelated? Of course, part of the reason for the polarisation of opinion is the very human tendency for disagreement. It is as though the process of scholarship itself, and nowadays academic survival, depends on defence and attack. Evolution, because it deals with a past that can never really be known, was once ideal material for perpetual debate. Enter genetics with a completely new source of objective data. Surely, these old questions would soon be settled one way or the other. Or would they?

I had the opportunity to assemble eight contemporary opinions on this question thanks to the annual Wolfson College lectures. The eight weekly talks, held in the Spring of 1997, played to packed houses. I originally had no plans to publish the lectures, largely because I know from my own experience that giving a lecture is one thing and delivering a manuscript quite another. However, the list of speakers was so exceptional and the talks so very good that I changed my mind. To their credit, Oxford University Press accepted the book proposal with admirable speed and this helped persuade the contributors who, working from transcripts, produced some excellent essays which required only the lightest editing.

We begin with Colin Renfrew's wide-ranging introduction, which encapsulates the idea of a great synthesis of genes, language, and archaeology. Somehow, one feels instinctively, they must all fit together into the single true history of human populations. But how? He touches on a key point—the difference between a phenogram (a diagram of overall similarity between populations or languages) and a true evolutionary tree. On the face of it this might seem to be a subtle irrelevance, but, as subsequent chapters reveal, it actually highlights a stark division between two schools of thought. First, in language between adherents to Professor Joseph Greenberg's well publicized views and the body of historical linguists for whom Don Ringe's essay is a well argued manifesto. The same point also lies at the root of the argument between Professor Luca Cavalli-Sforza whose pioneering work on classical genetic markers established the field of human population genetics almost single-handed, and ingenues like myself who work with genes like mitochondria whose detailed evolution can be traced. Chris Stringer is also no stranger to controversy though his side has all but won the argument as to whether modern humans emerged relatively recently from Africa or—a view energetically defended by Professor Milford Wolpoff and his supporters—evolved from earlier waves of migration of *Homo erectus*. In his contribution, Chris compares the palaeontological evidence for this transition in Europe and Australasia.

In July 1997, after his lecture but before this book appeared, Svante Pääbo published what is arguably the most important paper thus far in the field of ancient DNA—the recovery of mitochondrial sequences from the Neanderthal type specimen. Several contributors refer to this paper in their written contributions, but, because it was a closely guarded secret before publication, they could not do so in the lectures. In my edited version of his lecture, you can see he is giving nothing away and is positively downbeat about the prospects for ancient DNA in modern humans. Of course, he knew by then that Neanderthals were not modern humans, but that didn't stop people who had heard him speak from being surprised at the turn of events when they heard the news of this success a few months later.

The lecture and the book gave Gabby Dover a good reason to gather his thoughts on the question of 'genes' for language, consciousness and other attributes that set humans aside from other animals. The result is a delightful and original essay where he reveals his thesis that it was all done, as he puts it, by 'teaching old genes new tricks'.

Finally, no series on the genetics of human populations would be complete without Walter Bodmer. Recently returned to Oxford as Principal of Hertford College, he has been a close colleague of Professor Cavalli-Sforza for many years and their early books were the standard texts for me and other younger geneticists. A vociferous advocate of the phenetic school of population genetics, he rounds off the book, as he did the lecture series, with a review of the contributions from HLA, his favourite classical marker, and a timely reminder that we neglect natural selection at our peril.

My brief to the speakers was to be opinionated and dogmatic. I quickly abandoned thoughts of there being any sort of consensus at the end. Instead, we have a book of essays, presented in the order they were given. Not, by any stretch of the imagination, a balanced selection putting all sides of the argument, but eight contemporary opinions which do not attempt to disguise the tensions in the field.

As well as the contributors, there are others without whom this book could not have appeared. My grateful thanks to the President of Wolfson College, Sir David Smith, for presiding over the lectures, to John Penney for his encouragement and help with assembling the programme, to Jan Scriven for arranging for recordings and transcripts and, especially, to Pat French for her tireless help with the manuscripts.

Oxford   B. S.
December 1998

Contents

Contributors

Sir Walter Bodmer FRS
Principal, Hertford College, Oxford

Professor Gabriel Dover
Professor of Genetics, University of Leicester

Professor Svante Pääbo
Professor of General Biology, Zoologisches Institut Munich

Professor Lord Renfrew of Kaimsthorn FBA
Disney Professor of Archaeology, University of Cambridge

Professor Don Ringe
Professor of Linguistics, University of Pennsylvania

Professor Chris Stringer
Department of Palaeontology, Natural History Museum, London

Professor Bryan Sykes
Professor of Human Genetics, University of Oxford

Professor Ryk Ward
Professor of Biological Anthropology, University of Oxford

1

Reflections on the archaeology of linguistic diversity

Colin Renfrew

1.1   A NEW SYNTHESIS?

We are, of course, as humans all members of one species, *Homo sapiens sapiens*, and the differences among us are really quite limited. Many of these are acquired cultural differences. These are the things which we learn and the various distinctions between us which develop after our birth. Language ability seems to be a general human attribute but the specific language which we learn in childhood is one of these acquired characteristics of the human individual. The specific culture in which we participate is another. Other differences are, of course, genetically determined.

I am an archaeologist and make no claim to be a historical linguist, so you may find it rather paradoxical that I am trying to write about world linguistic diversity. Certainly the fact that I am not a historical linguist would be warmly confirmed by my friends in that discipline and indeed others in that discipline also! It is the case that we may be on the brink of a new synthesis as is indeed foreshadowed by some of the contributions in this volume. They deal with linguistics, with genetics, with evolution, and with human diversity. We may be on the brink of seeing some convergence in our understanding of issues of genetic diversity, cultural diversity, and linguistic diversity.

It may be possible, then, to work towards a unified reconstruction of the history of human populations. It is much needed, because certainly we do not have such a unified history at the moment. There is nothing in historical linguistics, as currently conceived, which meshes in any coherent way with the evolution of our own species or with what we understand of evolution within our own species. And there is very little in historical linguistics that correlates in any significant way with what we understand of the human past through archaeology, that is to say, through prehistoric archaeology. Most linguists hold the view that there is nothing much that you can say beyond a barrier of about 6–8000 years ago so far as language is concerned. There is much interest at the moment in seeing if this received wisdom is indeed valid.

If it is, then we are in for a disappointment and there won't be any great new synthesis, because the issues which many of us are interested in as archaeologists, as molecular geneticists and as evolutionary biologists, go well back beyond 6000 BC. So the real question that faces us and which, if I may be so bold as to say so, you might find interesting to bear in mind throughout this book is this: can we really get these different categories of evidence—the genetic, the cultural/archaeological, the linguistic, the biological/anthropological—to mesh and link up with each other?

Archaeology is good at time depth. We can date artefacts left by earlier peoples, but that doesn't tell us in itself very much about their genetic makeup and certainly tells us nothing about their language. Those are a matter of inference. Historical linguistics is very good at comparing living languages and indeed very good at dealing with written languages. But written languages do not go back earlier than 3000 BC and historical linguistics, not surprisingly, becomes enmeshed in uncertainties when one goes back much before that time. Historical linguistics has real problems, in my estimation, with time depth. Sometimes claims are put forward that you can make statements about the date when one language diverged from another, maybe 4000 years ago. I am sceptical about the chronological basis.

Molecular genetics is good at investigating living populations and most of the statements emerging today from molecular genetics about early humans are based on analyses from living populations. One very interesting field which is emerging, and one which will be covered in Chapter 6, is the initiative to get molecular genetic information from early human material, and thereby get direct data of a molecular genetic nature from ancient hominid remains. So far that hasn't really advanced very markedly, although the recent success in analysing samples from Neanderthal man[1] gives an early indication of what may one day be achieved. Molecular genetics is another discipline which finds it quite difficult to make accurate statements about the past, or rather about the chronology of the past using contemporary data as the starting point. I will return to that point in a moment.

Much of the interest of the current situation comes from attempts by linguists to cope with time depth. There are great controversies in contemporary linguistics about how far back you can go. The concept of the language family is well established but can you go back beyond the language family to more embracing and earlier entities, the so-called macro families? That is a matter which is hotly contested. I think it is fair to say that molecular geneticists, in their attempts to cope with time depth, also have their problems. There has been a considerable fiasco recently with the interpretation of mitochondrial DNA, supporting the so-called 'Out of Africa' argument for the origins of the human species. Very firm claims were made by Cann, Stoneking and Wilson on the basis of reconstructed classificatory trees.[1] But it then turned out that the reconstruction of those trees could be bettered using slightly different methods: quantitative statistical methods, which gave a completely different answer. There was a certain amount of embarrassment about that and a certain amount of re-structuring of maximum parsimony trees. Then, lo and behold, some even more parsimonious trees were found that again yielded the original answer, the 'Out of Africa' answer, which is, of course, supported by other genetic evidence. As a scientific demonstration it was not very convincing. I am not criticising the quality of the analytical work but the quality of the inference did suffer a bit of a blip, I don't think it's unfair to say.

The aim of Fig. 1.1 is to remind you of the components of what a first synthesis may yet become. It is very difficult to demonstrate, by means of a map, what the extent of linguistic diversity may be.

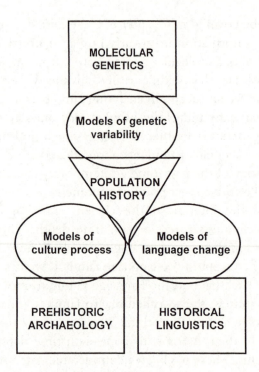

Fig. 1.1 The New Synthesis? The relationship between molecular genetics, prehistoric archaeology and historical linguistics.

In the case of South America, for instance, linguists have prepared maps showing the language families on a generally accepted classification and they yield a picture of very considerable complexity. And of course when you look at a language map of the world, it is very difficult to reduce to a coherent order; you can only do so by very grossly oversimplifying it into a limited number of categories.

If we turn from linguistic to cultural diversity, it seems at first sight easier to discern some outlines. Hunter-gatherer economies survived into the present century but only in lands which we would today consider as economically marginal and in general unsuitable for the exploitation of domesticated plants and animals. In the Near East there was one influential growth trajectory, developing first in the Fertile Crescent, which set the pattern for development in that area and subsequently in Europe. The trajectory in China is a very different one. In South America and in Central America again very different patterns of development are seen. In each area there is a good deal of local diversity, where adjacent groups can have very different life styles. For instance, the nomad pastoralists of Iran exploit some of the same regions as the settled farmers of the area, but in a very different way. Again and again we see communities with special features peculiar to themselves.

When we look at contemporary genetic diversity it is helpful to remember that the gene frequency distributions of the world show a great deal of geographical variety for individual genes, and I think most geneticists would agree today that the old concept of the 'races' of humankind is a very gross over-simplification. It is just one way of describing this considerable degree of diversity. For one thing, people who used to speak about race were emphasising only one group of features, generally features relating to pigmentation of skin and hair and eyes, and to some extent features of the face. But now that we have gene frequency maps for so many genes, many of which deal with perhaps rather more fundamental attributes, it is clear that the old method of summarising the races of man into five or six superficial categories is not a very effective way of reviewing the overall picture. It seems clear that genetic drift plays an important role. And when you have continents or sub-continents that are remote one from another, such as Australia, then you get isolation and divergence, and hence the development of localised features.

Now let us now turn to language. Charles Darwin made the following statement:

If we possessed a perfect pedigree of mankind, a genealogical arrangement of the races of man would afford the best classification of the various languages now spoken throughout the world; and if all the extinct languages, and all intermediate and slowly changing dialects had to be included, such an arrangement would, I think, be the only possible one . . . this would be strictly natural, as it would connect together all languages, extinct and modern, by the closest affinities, and would give the filiation and origin of each tongue. [2]

This statement looks quite innocent but it does, in fact, introduce a series of important and perhaps novel concepts which slide in there without your noticing. He speaks of 'pedigree' and that introduces the concept of the family tree, an image which underlies so much discussion about evolution. It introduces the concept of descent, and when you talk of descent you are also very close to talking about divergence. If you look at any tree model for anything, you're talking about divergence and you very soon have in mind concepts such as what geneticists today would call 'genetic drift'. Well, language has its own processes of divergence whereby languages in isolation move further and further apart by a process very closely related to genetic drift.

It was Thomas Huxley who introduced the essential second idea. Huxley pointed out that languages can be acquired by means other than genetic:

It seems to me obvious that, though in the absence of any evidence to the contrary, unity of languages may afford a certain presumption in favour of the unity of stock of the peoples speaking those languages, it cannot be held to prove that unity of stock, unless philologers are prepared to demonstrate that no nation can lose its language and acquire that of a distinct nation without a change of blood corresponding with the change of language.[3]

This is an obvious enough point. It introduces the idea of language replacement within an existing population. That is why cultural evolution and indeed linguistic evolution proceeds in ways sometimes very different from genetic evolution. There are also possibilities of convergence, of interaction, of gene flow, if you like that image. Another series of questions that we need to think about is the question of rates of change. The first tree that was thought about extensively in such terms was probably that of the Romance languages. Quite early on in the history of linguistics, before the eighteenth century, people were well aware that many of what we today regard as the Romance languages were descended from Latin. I think this particular linguistic story, the descent of many of the languages of contemporary Europe from Latin, underlies much subsequent thinking.

1.2   THE HISTORICAL CONTEXT

If we look at the framework within which we are working, it is largely known and accepted today that our species, *Homo sapiens*

sapiens, evolved from an earlier species, *Homo erectus*, which itself evolved probably in Africa from anterior species, notably *Australopithecus*. We have a reliable time depth for these, obtained by radiometric dating methods, and it is generally agreed that our own species emerged from *Homo erectus* between 200 000 and 100 000 years ago. It is thought by many that this transition of our own species, *H. sapiens sapiens*, took place in Africa, although that point is still a subject for argument and discussion, and one which Professor Stringer considers in Chapter 2. In Fig. 1.2 you have the two alternatives. On the left, you have the notion that there was a much earlier origin for *H. sapiens sapiens*, emerging from *H. erectus* in South-east Asia, in Africa and perhaps in Europe, no doubt with some interactions between them. The alternative, favoured by most anthropologists, and shown on the right, is that the other lineages of *Homo erectus* in South-east Asia and elsewhere became extinct and were replaced by a new radiation of hominids, this

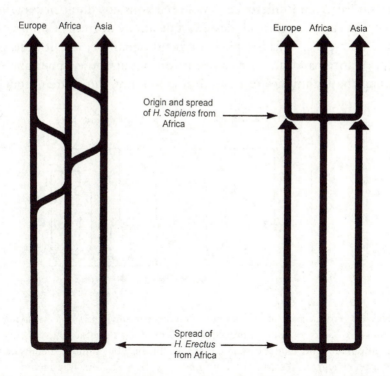

Fig. 1.2 Alternative models of human evolution ('multilineal' vs 'Out of Africa').

time *H. sapiens sapiens*, out of Africa around 100 000 years ago. But this is still a lively controversy. For instance, scholars in China are persuaded that their *H. erectus* is ancestral to their own *H. sapiens sapiens* populations.

On Fig. 1.3, you see what is, to some extent, one of the nubs of contemporary arguments. I indicated previously that the linguistic evidence is mainly contemporary, and one of the purposes of historical linguists is to reconstruct the ancestral language, or proto-language, and in so doing to work back into the past using data, either contemporary or with a limited time depth. That, too, has been until recently, and to a large extent still is, the approach of molecular geneticists who take samples from living populations and try, on the basis of similarity and difference between those samples, to say something about the likely nature of ancestral populations. On the left of Fig. 1.3 is a phenetic dendrogram, where one is looking at modern populations and arranging them in terms of similarity and difference. These relationships don't necessarily tell you about origins and descent, but simply degrees of similarity and difference. This is a phenetic dendrogram because it deals in appearances and arranges information about those appearances, using the techniques of numerical taxonomy, into a tree model.

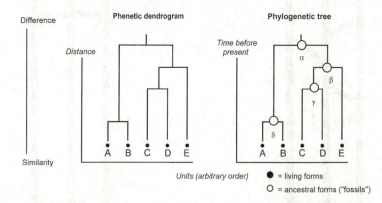

Fig. 1.3 From classification (taxonomy) to descent (phylogeny). In classifying a number of units a common procedure is to produce a tree diagram (phenetic dendrogram: left). Under certain conditions this may be considered equivalent to a phylogenetic tree (right) whose nodes (α, β, γ etc.) may be recorded as fossil forms.

But it does not need to be the same as a phylogenetic tree. A phylogenetic tree is a real pedigree, in a Darwinian sense, and does give you descent. The nodes on the tree have real historical meaning. They really existed. Most of the difficulties and problems in the interpretations of historical linguistics and molecular genetics arise from the difficulties in reaching a plausible phylogenetic tree, a real family tree diagram operating in real time, using data which are collected today from recent languages or from contemporary samples from different populations. That is a point to which we will return on a number of occasions.

DNA, nuclear DNA, has been used in roughly that way in the work of Clegg and colleagues in Oxford,[4] using beta-globin gene clusters to look at different human populations, to see how they are related and to suggest that there may well some sort of family tree construction (Fig. 1.4). Since Africa branches off first from the rest in the phylogenetic tree, that would support an 'Out of Africa' origin for our species.

It is important to recognise that much of the information that is discussed these days comes not from the nucleus, but from mitochondria. These are energy producing elements within the cell, but not within the nucleus of the cell, and which are simpler than the nucleus. They have about 16 000 base-pairs compared with the 3 000 000 000 base-pairs of the nucleus, and they also have the remarkable property of being passed on exclusively in the maternal line. So your mitochondria resemble those of your maternal grandmother and of her maternal grandmother and so forth. If one

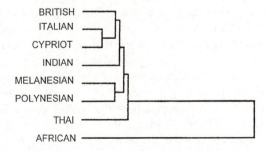

Fig. 1.4 Dendrogram based on data from the beta-globin gene cluster indicating the relationships between seven modern human populations. (Redrawn from Wainscoat et al.[4])

looks at the data from mitochondrial DNA, despite the slight blip in interpretation which I mentioned earlier, they are still largely interpreted in much the same way using a phylogenetic tree diagram. In Fig. 1.5 it has been bent round for convenience in putting it on the page. The first division is again exclusively African and then you have a whole lot of other things happening in the very complicated history which one is naturally involved with. In both these cases, modern samples—these are samples taken from living populations—have been used.

1.3   LANGUAGE

I shall now move on to languages and ask what are the main units of linguistic classification. The first unit is naturally the language itself. Most of us have some notion of what a language is and how you distinguish between a language and a mere dialect. It was Sir William Jones who, in the year 1786, made a very significant discovery, a discovery about what we now call the Indo-European languages; this was also one of the first recognitions of the concept of a language family. It is this concept which I want to emphasise here. It is still problematic, though it has been a central concept in linguistics for two hundred years. Jones worked in India as a judge and, in his famous address to the Asiatic Society of Bengal he remarked as follows:

The Sanskrit language, whatever may be its antiquity, is of a wonderful structure, more perfect than the Greek, more copious than the Latin, and more exquisitely refined than either, yet bearing to both of them a stronger affinity, both in the roots of verbs and in the forms of grammar than could possibly have been produced by accident; so strong indeed that no philologer could examine them all three, without believing them to have sprung from some common source, which perhaps no longer exists; there is a similar reason, though not quite so forcible, for supposing that both the Gothic and the Celtic, though blended with a very different idiom, had the same origin with the Sanskit; and the old Persian might be added to the same family.[5]

Here we see Jones doing a number of things. He is asserting that these languages are related in a language family. He is asserting

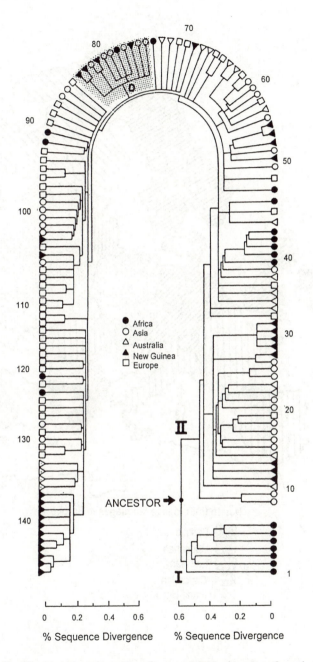

Fig. 1.5 Dendrogram based on data from mtDNA showing the relationship between 147 modern human individuals and indicating that the first significant branching is between Africans and non-Africans. (Redrawn from Cann *et al.*[1])

Fig. 1.6 The principal Indo–European language groups of Europe and Asia indicated by shading. Some extinct Indo–European languages (Hittite, Tocharian) are also marked.

Indo-European languages in Europe and Asia

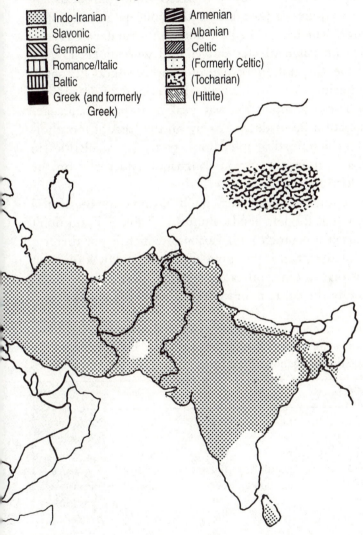

that, therefore, they must have sprung from some common source. He is thereby constructing the notion, essential to historical linguistics, of the proto-language, the notional ancestral language of a language family which is the progenitor of all the relevant languages currently spoken or spoken in recent times. Later on this particular family was called 'Indo-European' by scholars and so Jones is introducing the concept of the Indo-European proto-language. Then the question naturally arises as to where this proto-language was spoken and who spoke it. Very soon you have the notion of a proto- or pre-Indo-European speaking population and perhaps also the notion of an original homeland where that population was speaking its proto-language.

The observation he made was partly on the basis of morphology, including grammatical inflections, partly by regularities in sound changes among the various languages, and partly on the study of the lexicon, the vocabulary.

The basic numerals offer a very good example for the lexical approach. Look at the left-hand columns in Table 1.1 and don't concern yourself too much with Japanese, which is put there to show how different it is. If you ponder that list just by looking at the numerals one to ten, you can judge that there is some commonality among the contemporary Indo-European languages and

Table 1.1 The numbers one to ten from five Indo-European languages and from Japanese reveal at once, upon inspection, the affinities of the former and the distinctiveness of the latter.

English	Old German	Latin	Greek	Sanskrit	Japanese
one	ains	unus	heis	ekas	hiitotsu
two	twai	duo	duo	dva	futatsu
three	thrija	tres	treis	tryas	mittsu
four	fidwor	quattuor	tettares	catvaras	yottsu
five	fimf	quinque	pente	panca	itsutsu
six	saihs	sex	heks	sat	muttsu
seven	sibum	septem	hepta	sapta	nanatsu
eight	ahtau	octo	okto	asta	yattsu
nine	nium	novem	ennea	nava	kokonotsu
ten	taihum	decem	deka	dasa	to

that perhaps they have sprung from some common source. It's a very limited data-set, just ten words. Jones also considered phonology, sound change, but I don't want to go into that now.

Figure 1.6 shows the very wide distribution in Europe and Asia of the Indo-European languages. Nobody much doubts, apart from just one or two scholars, that they have all indeed sprung from some common linguistic source. Very possibly, there was a group of people, somewhere, who spoke proto-Indo-European, and the languages of Europe and the relevant parts of Asia may indeed derive from them and from their speech. It is worth emphasising the interest of looking at the non-Indo-European languages: the Basque language in north Spain; perhaps Pictish in Scotland, now extinct; Etruscan in Italy, also now extinct; the Caucasian languages, very different from Indo-European; and the Finno-Ugrian family, including Hungarian. Turkish, of course, belongs to a different language family, but in earlier times there were Indo-European languages in Anatolia. So here we have this vast tract of land. How on earth do we explain this wide distribution in historical terms?

The first approach, very soon after Darwin's writings and still in the 1860s, was Schleicher's Tree (Fig. 1.7), a simple family tree, which, I think, reflects in a way the insights of Sir William Jones almost a century earlier. But it is worth mentioning that other models have been applied. A wave model, which perhaps moves away from the notion of having a single tree, with innovations causing divergence, is one. Here innovations are seen spreading through the whole area, quite possibly from different points of origin within a wide region. The wave model does open the way to thoughts of convergence.

There was one linguist who, very heretically, suggested that there might not have been a single ancestral proto-Indo-European language, but rather that those populations who came to speak Indo-European did so through a process of contact, of learning each other's languages, a process of convergence. That was Count Trubetskoy,[6] but Count Trubetskoy does not get a good press in this respect from most linguists today. And yet at the same time, convergence processes are interesting and many linguists today

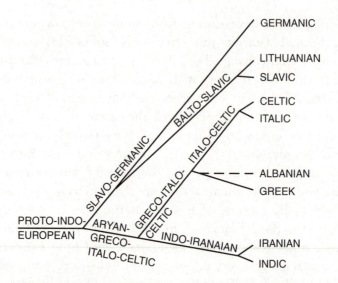

Fig. 1.7 The tree model for both the classification and the descent of languages in the Indo-European family, as first drawn by Augustus Schleicher in 1863 under the influence of Darwin.

discuss the formation of Creole languages, which carries something of the same idea.

Can one use the historical method, the comparative method, to take the idea of language families a little further? If we compare languages and establish a language family, is it possible to compare different language families and perhaps reach some notion of a macro-family? That is an initiative that has been undertaken by some scholars and has proved very controversial. Joseph Greenberg classified the languages of Africa into just four main major macro-families, and his classification for Africa has been widely accepted.[7] For the Americas, he has classified the American languages into Na Dene, Amerind, and Eskimo-Aleut[8] and has been soundly berated for his efforts. I don't propose to follow this classification by Greenberg and his colleague Merritt Ruhlen[9] here but simply indicate that it exists. It implies lumping a whole lot of American languages together as 'Amerind', and the same is true for certain other groups; for instance 'Indo-Pacific', formed by lumping the very different language families of New Guinea together into one group. But in doing that, I'm not suggesting that they are necessarily related genetically and belong to some larger unity.

1.4 APPROACHES TO LANGUAGE CHANGE

How are we going to address language change? I would suggest that there are four classes of model for language change. It is worth making these explicit and I think most historical linguists are speaking about change operating in one of these ways.

First of all, there is *initial colonisation*. You have an uninhabited land, people move in speaking a language and, in this way, a language is introduced. Secondly, and very much favoured as an explanation by earlier generations of scholars, is *language replacement*. Here either a new population moves in, or through whatever other means, the language spoken in a particular area is replaced by another language. That is a process which can be Darwinian or, as Huxley pointed out, can be non-Darwinian if the genetic composition of the pre-existing population remains unchanged. Furthermore, you have *divergence* models which are the most important models for language change in historical linguistics. And then you have the possibilities of *convergence* models, which are used by linguists to some extent. They talk of Sprachbund[2] effects, where you do have relationships among geographically adjacent languages that are the result of convergence, but I don't know of any historical linguists today that use convergence models actually to explain the origins of whole language families, though it is certainly an argument that could be adduced.

It is worth referring briefly to the different models for replacement which one might put forward. One of those might be the demography/subsistence model where you have some economic change which allows one group of people to become predominant in a territory, and really permits their language to prevail, mainly by force of numbers. That's where the demography comes in and the subsistence element is perhaps needed for them to do so. Secondly, there is the *lingua franca* phenomenon, whereby you have one language coming to be spoken in the area perhaps for trade reasons, as a second language, then becoming a first language for a younger generation. This process has been recognised as the starting point for some Creole languages. Thirdly, elite dominance, when a small minority moves in, takes over, and by their social dominance their language becomes prevalent. And then finally, system collapse. Where you have great empires, for instance the Roman empire, when central power collapses, you very often find demographic shifts and elite dominance effects as well which result in significant language changes.

The traditional approach to the Indo-European problem has been to see the solution in terms of elite dominance; to assume that there was an Indo-European homeland, very possibly in what is now the Ukraine. It has been suggested that the domestication of the horse was the key that gave these people the impetus to move west, and also later east to India, and so on. These, it is claimed, were warrior nomad horsemen who conquered Europe by a process of elite dominance perhaps somewhere around 3000 BC. This is the idea initially propagated by Gordon Childe[10] and embraced energetically by Professor Marija Gimbutas.[11] In my book, *Archaeology and Language*,[12] I discounted this explanation, one reason being that we have absolutely no evidence that the horse was ridden in Europe for military purposes prior to the first millennium BC. We have clear evidence that the horse was used to pull chariots, mainly for military purposes, from around 1500 BC in Europe and perhaps a little earlier in the steppelands. The notion that around 3000 BC there were warrior horsemen is simply not attested archaeologically. Now we could have quite an argument about that, but I'll make that assertion and move on.

I asked myself what, then, was a sufficiently important change in prehistoric Europe that it might have been associated with the spread of Indo-European? And the answer that I came to, the fundamental change in European prehistory, is the spread of agriculture. We know that agriculture originated in and around the broad area of the Near East called the fertile crescent because it is there that the wild plants and animals which later became domesticated were originally found. If you look at archaeological sites in the area, famous sites such as Ali Kosh[3] and Jericho, and Çatal Hüyük and Hacilar in Anatolia—you do indeed find, in 6000 BC or earlier, some of the earliest domesticated plants. So it's clear that the domesticates which were the basis for farming did indeed spread from the Near East, including Anatolia, across to Europe. I think nobody today doubts that. The question is what were the consequences of that spread?

I suggested that it was that spread of people, as well as plants and animals from Anatolia to Europe, which brought the proto-Indo-European language, first of all to Greece and then to the Balkans, also to the west Mediterranean, central Europe, France, Britain, and so forth. And I did indeed suggest that we could link this idea with the model developed by Albert Ammerman and Luca Cavalli-Sforza[13] that there was a significant spread of people. The individuals did not actually move very far, perhaps twenty or thirty kilometres for any one person, but because of the great growth of population associated with the spread of farming which Ammerman and Cavalli- Sforza emphasised, you had the dynamic that produced what has been called a 'wave of advance', which in the course of a thousand or more years brought farming across Europe. Thus there developed a new farming population and the new language or languages which we could describe as proto-Indo-European. So that was what I suggested at that time. Others have emphasised, quite rightly, that the Mesolithic populations of Europe should not be overlooked. Marek Zvelebil[14] and others have accepted my model for south-east Europe and the middle Danube, but then suggested that further west you have to look at the interactions between the Mesolithic populations, the already resident populations, and the farming population that has come in (see Fig. 1.8). You should give due attention to the local populations, some of whose languages survive, like Basque and Pictish and Etruscan, which were mentioned earlier. So, that's one approach to the issue, though as we will see later, there are other approaches which may yet prove equally effective.

I want to pick up this idea of agricultural dispersal as a more general phenomenon, but I want to make the point that, although one simple form of the agricultural dispersal model will have the

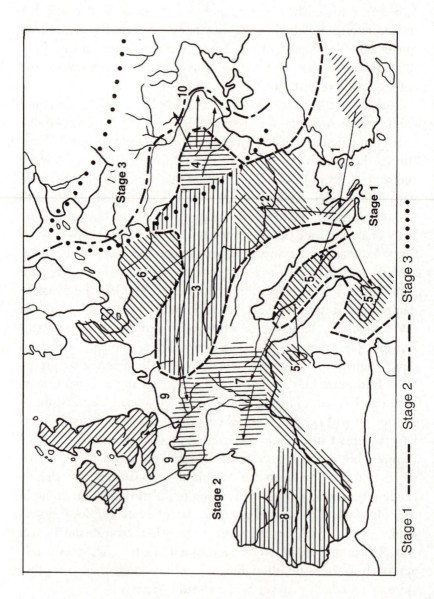

Fig. 1.8 Spread of Indo-European languages as suggested by Zvelebil and Zvelebil,[14] modifying Renfrew.[12] Hatched areas denote prehistoric cultures in Europe: 1. Early Greek Neolithic; 2. Stacevo/Körös/Karanovo; 3. Linear Pottery Culture; 4. Proto-Cucuteni and Proto-Tripolye; 5. Impressed Ware cultures; 6. TRB cultures; 7. French Neolithic; 8. Iberian Neolithic; 9. Neolithic of Britain and north-west European littoral. The three stages of dispersal are superimposed on this pattern. Stage 1: Introduction of agro-pastoral farming 6500–5000 BC; Stage 2: Consolidation of farming and 'secondary products revolution' 4800–2500 BC; Stage 3: Elite dominance 3800–1000 BC.

dispersal accompanied by an expanding population, so that there will be very strong genetic and linguistic implications, it is possible to devise alternative models. The linguist Ehret has spoken about such cases, where you have an economy such as agriculture spreading through interaction and producing a linguistic impact, so that you find languages spreading without any significant movement of population.[15] In such a case, you have linguistic change without very significant genetic change: what one might describe as contact-induced language change. The point is that when farming methods are devised in one area, they are, of their nature, inherently expansive. And that, I think, is borne out again and again in world prehistory. This expansive quality is very often reflected in language spread as well as in spread of agriculture. That is widely accepted now for the spread of the Bantu languages in Africa, the Niger-Kordofanian language family, and archaeologists claim to be able to trace that spread and to identify significant agricultural introductions.

The same point has been made with great force by Peter Bellwood for the languages of the Pacific, not just the Polynesian languages but the wider language family, Austronesian,[16] and Charles Higham has also argued along similar lines for South-east Asia.[17] You could imagine such a process taking place elsewhere and being associated with linguistic spread. As well as Indo-European you could make the case for Afro-Asiatic, and perhaps for other language families such as Elamo-Dravidian. That would, of course, be very hypothetical, but for those who are enthused about the Nostratic hypothesis, which sees all these languages as part of a single macro-family, or about Greenberg's Eurasiatic hypothesis, it might give some insight into the original relationship between those languages.

Perhaps it is possible to consider the languages of the world using this sort of approach. We badly need some overall understanding of how the languages of the world come to have the distributions which they do have. The linguist Johanna Nichols has suggested that you can see two kinds of language distributions.[18] One she calls 'spread zones', the other 'residual zones'. I suggest that you can indeed see that there are some areas of the world where the languages have not been affected by processual phenomena more recent than the initial dispersal. In those areas there has been a very long time since initial population for divergence processes to occur and for the local languages to become very different from one another, so there is a great profusion of language families. These are the residual zones. The Caucasus is one such area, New Guinea is another and north Australia is a third. But such effects in other parts of the world have been completely masked or swept away by spread phenomena, mainly of agricultural dispersal, but perhaps others as well.

Using these concepts, and accepting the 'Out of Africa' model for human origins, we can suggest a rather simple outline story for the origins of world linguistic diversity (Fig. 1.9). I suggest that one may think of four main models, or four principal processes at work. First of all, initial migration would apply to much of the world and we should note on the map (Fig. 1.9) those areas (remembering that these are contemporary languages) where the populations may have become established already long ago by a process of initial migration

---

Fig. 1.9 Modern distribution of world language macrofamilies and language areas as summarised by Ruhlen.[9] 1. Khoisan; 2. Niger-Kordofanian; 3. Nilo-Saharan; 4. Afro-Asiatic; 5. Caucasian; 6. Indo-European; 7. Uralic Yukaghir; 8. Altaic; 9. Chukchi-Kamchatkan; 10. Eskimo-Aleut; 11. Elamo-Dravidian; 12. Sino-Tibetan; 13. 'Austric'; 14. 'Indo-Pacific'; 15. Australian; 16. Na-Dene' 17. 'Amerind'. (Austronesian unshaded).

Note that while some of these families and macrofamilies are generally accepted as valid entities others (notably numbers 8, 13, 14 and 17) should perhaps be taken rather as denoting language areas, without implying the hypothesis that the languages within each are genetically related.

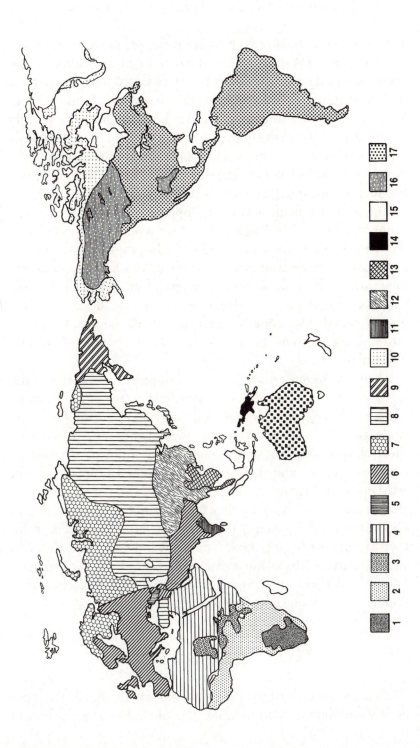

and have only been affected by subsequent spread phenomena to a limited extent. Those are the areas where language divergence has been continuing and where you might expect to have a very great deal of linguistic diversity because they have been in place for so long—the Khoisan languages, the Australian languages, especially the northern Australian languages, perhaps also in South-east Asia and New Guinea, and this would apply to most of the languages of North and South America, whether or not we see them as linked together by Greenberg's concept of the 'Amerind' macrofamily.

The second process is farming dispersal, which I have spoken about sufficiently. The languages we have spoken about remain on the map. To them must be added those languages whose primary distribution is to be attributed to the spread of a farming economy: they include the Indo-European languages, the Elamo-Dravidian languages, the Afro-Asiatic languages (including the Semitic languages), and the Austronesian languages, all of which took up their distributions, it could be argued, mainly as a result of farming dispersal processes.

Then you have late climate-related dispersals at the end of the Ice Age. These are mainly the northerly language groups: Finno-Ugrian, Chukchi-Kamchatkan, Eskimo-Aleut and the Na Dene languages. These must have been very significantly affected by those climatic changes, and somehow they must have taken up their positions around that time through population adjustments early in the Holocene period.

That leaves you with a rather limited amount of explaining to do, in terms of the fourth process, elite dominance. The main cases are the Altaic languages, which cover a very wide area, and also the arrival of Indo-European languages in India and Pakistan. I don't think there are many other very large scale instances of elite dominance until one comes to the colonial period.

1.5   THE GENETIC EVIDENCE

Well, those are rather bold proposals, but where does genetics get us? I would like now to indicate, first of all, that there is indeed some support from molecular genetics for some of these ideas, although I'm sure it's too early to come to cut-and-dried conclusions. I am certainly not suggesting that I'm offering the new synthesis here and now—quite the contrary.

Of course, genetics has nothing to say directly about linguistics, or vice versa. The link has to be the history of human populations, as I indicated at the outset. Molecular genetics can help us study the history of human populations. That does not in itself tell us about the history of languages. On the other hand, I think everybody would accept that the history of languages is inextricably linked with the history of populations. And what really makes life complicated is the phenomenon of elite dominance or those other processes by which languages can be overlain by other languages without very significant genetic change. As I have indicated earlier, you can think of models for the spread of farming where the human population was not greatly changed. Where there was no great influx of new farmers or new people but where farming techniques were taken up locally and where the impact, the cultural impact, was so strong that it might also have been associated with linguistic change: this may be termed contact-induced language change. So that is a possibility perhaps worth considering.

When Excoffier and colleagues looked at African populations sampled on the basis of tribal affinity, which in some cases is very much the same as linguistic affinity, they found a very strong correlation between genetics and language.[19] They looked at gamma-globulin[4] and produced a tree, a phenetic dendrogram, of similarities and differences on the basis of the observed genetics. They found that the structure they had arrived at correlated completely with the linguistic affiliation. The Afro-Asiatic languages are all grouped together in this method and there is also some grouping of other language families. So, in this case, if you do classify the populations in terms of the molecular genetics, you emerge at a classification that correlates with language.

Indeed, it is the case that if you look at world populations as a whole, language is the best predictor of genetic composition. You can think of many ways why it wouldn't be ideal in view of language replacement, but I think the statement is a true one, that language is the best predictor, a much better predictor than location. In Africa, for instance, you have different populations living quite close together and you can predict their genetic composition by their language affiliation much more accurately than you can by geographical location.

Turning now to Europe: Luca Cavalli-Sforza and colleagues[20] looked at classical genetic markers for Europe and did a principal components analysis (see Fig. 1.10). In that way, they were able to suggest that the first principal component gives strong support for their farming dispersal model, assuming that farming was having a significant genetic effect, and that the farmers were on the move. The clines, the strong directionality from south-east (i.e. Anatolia) to north-west support the notion of a gradual movement of population in that direction, as the farming dispersal model suggests. But here we come to the snag that exists with some genetic methods. The pattern can be explained in that way but there is

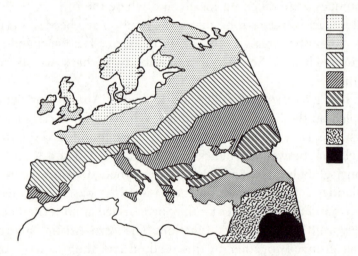

Fig. 1.10 Molecular genetic evidence for the spread of farming into Europe. The map illustrates the first principal component (28.1% of total variance) of molecular genetic variation in Europe, suggesting demic diffusion from Anatolia, probably associated with the spread of farming. The range between the maximum and minimum values of the PC has been divided into eight equal classes, which are indicated by different intensities of shading. The direction of increase of PC values is arbitrary. (After Cavalli-Sforza et al.,[20] p. 292.)

nothing inherent in the pattern to tell you anything about the timing. If you are sampling modern populations and looking at the first principal component, what you have in a way is a palimpsest of all time periods. Martin Richards and Bryan Sykes, in Oxford, have recently suggested, as I'll indicate in moment, that the pattern we are seeing here perhaps has much earlier origins going back to human dispersals in the Upper Palaeolithic period.[21] So these are the sort of pitfalls that you get when you try to match one category of data with another, but there is no doubt that the data do have a bearing.

If you look at the map of the frequency distribution of the rhesus-negative gene, a number of people have pointed out that in Europe you have particularly high frequency in north Spain. And this area is precisely where the Basque language is spoken, and the Basque language may be regarded as the only early and indigenous language of Europe. It's been there for some time; it is non-Indo-European, and is indeed apparently not closely related to any other language. Of course, elsewhere in Europe you have the Finno-Ugrian languages (including Hungarian), and also Turkish, but we now have some understanding that those are later arrivals. It is very possible that the Basque language was spoken in Europe by our early ancestors, before the Indo-European or proto-Indo-European languages came to Europe, whenever that may have been. It is fascinating that you get some image of that, some reflection of that, in the frequency distribution map for the Rhesus-negative gene today.

Another piece of evidence relating to what was said earlier comes from statistical work by Guido Barbujani and colleagues. They looked at the frequencies of classical genetic markers for the populations of Europe and the Near East. They examined my hypothesis that a demographic process, a wave of advance, carried not only the proto-Indo-European language to Europe but also the Afro-Asiatic languages to North Africa, perhaps Elamo-Dravidian to India, and so on.[22] I don't have time to discuss this work in detail but they did see clines of gene frequency which, or so they argued, do indeed support the dispersal hypothesis. So you do find support for the farming dispersal hypothesis for some of those language families. But again, that doesn't give any precise time depths. The Neolithic hypothesis offers one suitable dispersal which could have generated the patterning observed. But if you can think of another process that would give you similar migratory effects in those directions, such as the initial peopling of those areas by humans in Upper Palaeolithic times, that might be equally valid.

Again and again, we come to the problem of time depth in molecular genetics, and again and again we come to the problem of time depth in linguistics. I assure you that when some linguists say that we think this or that change must have happened five or six thousand years ago, they have no very effective means of calibration. That is quite clear, and many of the arguments introduced by some linguists in support of the chronologies which they offer are, in my view, circular. They say we can make an age estimate of about five thousand years because we believe something similar about some other language family. But if you ask why they believe this about the other language family the argument tends to take you, by a long way around, right back to the one you first thought of. I know that comment will irritate some of my linguist colleagues but that is sometimes what happens.

Figure 1.11 is from the work of Martin Richards and Bryan Sykes and their colleagues, of which there will be more in Chapter 5. It is a diagram relating to the mitochondrial DNA of a number of living, mainly European, populations. They have recognized a number of lineage groups in Europe and have, using ingenious, although I think slightly debatable, methods, sought to date these clusters and calculate their divergence times. They point out that lineage group 2A on their chronology might be related to the spread of agriculture. But they make the very important and interesting suggestion that many of the lineages in Europe should be earlier—the divergence times are earlier—and this suggests to them that we are talking about the genetic composition of the European populations being largely determined by dispersals in Upper Palaeolithic times rather than subsequently. So there you have a direct conflict of view between themselves and Cavalli-Sforza and his colleagues. I have heard some of Professor Cavalli-

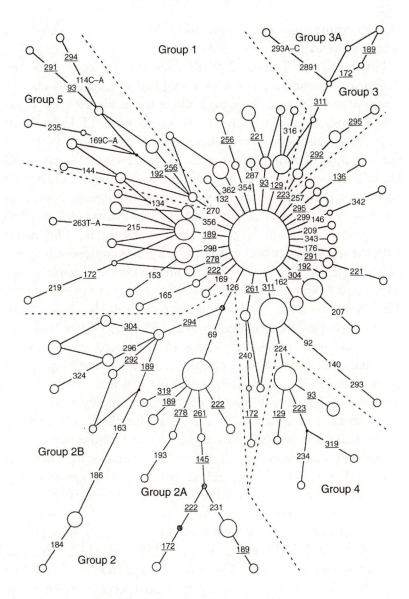

Fig. 1.11 The phylogenetic network of core European mitochondrial lineages. Circles are haplotypes with their areas proportional to the number of individuals who share it. Lines connecting circles are mutations in the mitochondrial control region whose positions are denoted by the numbers. Dotted lines demarcate putative clades or lineage groups. (After Richards et al.[21])

Sforza's comments on this work, and I think it would be a correct summary to say that he does not concur!

Finally I would like to allude to work that has gone on in the Americas. I mentioned that Greenberg had undertaken a classification of the American languages which has produced great vexation among American linguists. He speaks of the Eskimo-Aleut languages, and nobody particularly disputes that. Then he lumps the languages of various other groups together as the so-called Na Dene languages, and that isn't found particularly controversial either. But then he puts together all the other language families of the Americas, says they all have deeper resemblances and are all to be regarded as forming the 'Amerind' macrofamily. This meets with the great disapproval of the majority of Americanist linguists (see Chapter 3). In a way, it would not be very surprising if they were related, you might think, looking at the problem from afar. The Americas were indeed populated through some initial colonisation process, whether it was 15 000 or 35 000 years ago, so you would expect the populations to be related genetically, as indeed they turn out to be. You might also expect the languages to be related, but of course linguists in general feel that it is irresponsible to talk of such remote resemblances and such great time depths as these, and that the language families that one recognises are probably not related. This doesn't mean that the populations composing them may not ultimately be related some thousand years earlier, simply that to seek to compare languages at such a time depth is to go beyond the framework in which historical linguistics can operate, so those linguists say.

But that's not what Greenberg says. Greenberg says he thinks he can show that these languages are indeed related, and he has published his book to say so.[8] This is interesting, and I'm concluding with this point for two reasons. First of all, recent work on mitochondrial DNA by Torroni and his colleagues, has given evidence which they claim strongly supports the Greenberg hypothesis.[23] In other words, they find genetic relationship between the Na Dene speakers, genetic relationships, not surprisingly, between Eskimo-Aleut speakers, but also genetic relationships between the so-called Amerind speakers. Their interpretation lends strong support to the work of Greenberg. In Chapter 7 Professor Ryk Ward looks at the American languages, and it is my understanding that he feels that the situation is not so simple, and that perhaps the genetic evidence has to be differently interpreted. So I give you warning of that.

I'm deliberately finishing on this rather inconclusive note because the point of my discussion has not been to assert that the new synthesis is upon us. Although I have given you my own sort of grand outline for the origins of world linguistic diversity through the four processes I discussed, I think that there is a lot more that we have to learn. We have to learn much more from the geneticists and we shall, no doubt, do so. But we can expect a lot more also from the linguists because they, in my view, have to work further at refining and defining the notion of the language family and the possibility of macrofamilies. The majority of historical linguists would see it as a gross and unwarranted simplification to lump the world's language families into the few over-arching taxonomic units that we have been discussing here. There are, among linguists, lumpers and there are splitters. Of course, you can classify anything any way you want, but what we're talking about is a linguistic taxonomy, a phenetic dendrogram, which *might* actually have meaning in historical terms. There is one question that is really central to the understanding of evolution in these fields, in molecular genetics and in linguistics: when is a phenetic dendrogram a phylogenetic tree? I apologise for the jargon. If you don't understand it now I'm sorry, because it's been rather central to my discussion, but that is the question I would like to leave you with. When is a phenetic dendrogram a phylogenetic tree? If you can answer that question, you can probably write the new synthesis for yourself.

NOTES

1. The Neanderthal results were not published until after this lecture was given.

2. A Sprachbund ('language league') is a group of languages whose distributions are spatially adjacent but which are not necessarily genetically related, yet which come to show similarities both in vocabulary and in some phonological and morphological characteristics through the operation of convergence effects.
3. In what is now South-west Iran.
4. A blood protein associated with immunity. The genetic differences studied were based on protein not DNA variation.

2

The fossil record of the evolution of *Homo sapiens* in Europe and Australasia

Chris Stringer

2.1 INTRODUCTION

In this paper, I review the fossil evidence from the area which preserves the best record of recent human evolution, Europe, and compare it with the record from the other end of the inhabited world in the late Pleistocene, Australasia. I will also discuss the growing body of genetic data relevant to our origins. The debate about the tempo and mode of modern human evolution is still dominated by two contrasting models—that of Multiregional Evolution[1–3] and that of a single origin—the so-called 'Out of Africa' model.[4,5] Of course, both models posit an African origin for humans, but the former does so over a time scale of more than a million years (1 ma), while the latter does so over a time scale of less than 200 000 years (200 ka). Calibration is therefore of great importance in establishing the rate and manner of the establishment of the modern human characters in the different regions of early human occupation. Great progress has been made over the last decade in dating the fossil human record beyond the 30–40 000 year useable span of the radiocarbon method (e.g. references 6–8). Genetic methods are also, potentially, able to provide calibration points for past gene divergences, which can then be compared with dates for possible lineage splits in the fossil record. In both Europe and Australasia the apparent juxtaposition of late archaic humans and early modern humans about 40 000 years ago provides interesting tests of the competing models for modern human origins. In my view, the growing bodies of fossil and genetic evidence now make a global multiregional model covering a Pleistocene-long time-scale untenable, but an intermediate model which allows some degree of hybridisation between late archaic forms and early modern humans dispersing from Africa remains feasible.[9,10]

Europe and Australia are comparable in area, but while the former continent is really an extension of the much larger Asian land-mass, Australia is, and has always been, much more isolated. Both continents have been greatly affected by the periodic Pleistocene ice expansions, but while Europe suffered the very direct effects of ice in northern and upland areas, only the Tasmanian region of Australia was ever significantly glaciated, and instead, aridity was the main expression of the Ice Age in Australia. Humans adapted relatively early to periglacial conditions in Europe, in the form of the Neanderthals, whereas it seems that *Homo erectus* never reached Australia to encounter such environments.

The Neanderthals represented a successful lineage of early humans who adapted to a variety of European Pleistocene conditions—from glacial to interglacial, and from boreal to mediterranean.[11] They were highly evolved humans, both physically and culturally, and shared many features with us—for example, they had an extended growth period, small teeth and large brains. Their technology (termed Mousterian or Middle Palaeolithic) shared major features with that associated with the earlier stages of modern human evolution, and they evidently also showed advanced behaviours such as burial of the dead and social care of disabled individuals. During the final stages of their existence, the Neanderthals even showed a capability of manufacturing Upper Palaeolithic-style specialised artefacts of the kind once thought unique to modern humans.[12] Their geographical range as indicated by fossils extended from Britain and the Iberian peninsula in the west, to the Levant and Uzbekistan in the east. From artefactual evidence, it was no doubt even larger than this. However, no Neanderthals are yet known from Africa or the Far East.

I will discuss the likely antiquity of Neanderthal evolution later, but in terms of the whole time span of the genus *Homo*, which is about two million years, the Neanderthals, like us, came late in the story. The interface between the last Neanderthals and their successors in Europe, the Cro-Magnons, may have occurred as recently as 27 000 years ago.[13] Astonishingly, we are only separated from them by some 1500 generations! Practically every researcher, whatever their view of modern human origins, would accept that we and the Neanderthals ultimately derived from an African ancestral form represented by the material from Koobi Fora and Nariokotome (northern Kenya), attributed to the species *Homo ergaster* by some workers, or early *Homo erectus*, by others.[14] The marvellously preserved Nariokotome skeleton, dated to about 1.5 ma, illustrates that these early humans were fundamentally like us in their anatomy below the neck, although differing in details of their pelvis, vertebral column and rib cage. They had developed the basic human body shape characterised by relatively long legs and a proportionately smaller torso, compared with the earlier australopithecine physique, which was closer to that of the living great apes. Moreover, the tall, narrow-hipped physique of the Nariokotome boy is very comparable with that found in populations living in similar regions today—such a physique provides a high surface area/volume ratio, and is ideal for thermoregulation in hot, dry climates. But above the neck, these early humans were still quite distinct from us. Their brain sizes were about the same as that of a one-year-old human baby of today, although clearly exceeding those of the australopithecines, which barely reached the levels of even an average modern new-born. The brain was contained within a long and low braincase, and the face was broad and flat, surmounted by a large brow ridge, behind which was a receding forehead. The face jutted forwards, and this prognathism was accentuated by a chinless lower jaw.

2.2   THE EUROPEAN FOSSIL RECORD

It must have been populations like these which first emerged from their African homeland about 1.5–2 ma, and then spread, apparently rapidly, through subtropical regions into South-east Asia.

They may also have moved northwards to the eastern fringes of Europe, if the human mandible from Dmanisi in Georgia has been accurately dated to over 1.5 ma. The exact time of the first human colonisation of Europe is very much disputed, but the recent finds from Gran Dolina (Atapuerca, Spain) and Ceprano (Italy) may well indicate that humans were established at least in southern Europe by 700 ka.[15] The species represented is also unclear, with claims for the presence of *Homo erectus* itself (Ceprano) or a newly-identified species, *Homo antecessor* (Gran Dolina). The latter species is viewed by its proponents as being derived directly from *Homo ergaster* (not from *Homo erectus*, which therefore represents a separate lineage).[16]

By about 500 ka, the picture in Europe becomes a little clearer, with the appearance of a more derived species, generally termed *Homo heidelbergensis*.[15] The type specimen, a mandible, was discovered in the Mauer sandpit, near Heidelberg, in 1907. Subsequently, a number of other European and African fossils have also been attributed to this species, including finds from Petralona (Greece), Arago (France), Bilzingsleben (Germany), Broken Hill (Zambia), Elandsfontein (South Africa) and Bodo (Ethiopia). Most recently, a partial left tibia and two incisors from Boxgrove, West Sussex, have also been included. The Bodo fossils are perhaps the most ancient of these specimens, at about 600 ka, while Broken Hill, Petralona and Bilzingsleben may date to between about 300–400 ka. The limited amount of associated postcranial material precludes full skeletal reconstructions, but what evidence there is indicates a tall but very muscular physique, with possible adaptive differences already developing between tropical (Broken Hill) and temperate (Boxgrove) individuals. In my view, this species represented the last common ancestor of the separate lineages of Neanderthals (*Homo neanderthalensis*—in Europe) and modern humans (*Homo sapiens*—in Africa). However, the Spanish workers who have erected the species name *antecessor* for the more ancient Gran Dolina material instead argue that it is *this* species which lies at the root of the Neanderthal and modern human clades. By their reckoning, the succeeding *Homo heidelbergensis* was a European species only, one which was already part of the Neanderthal lineage. A separate and parallel line in Africa may have led through material like Elandsfontein and Broken Hill (*Homo rhodesiensis?*) towards *Homo sapiens*.[16]

In the Far East, it appears that *Homo erectus* still existed at about 300–400 ka, and as we shall see, there is evidence that it persisted for even longer in Australasia. By about 200 ka, more derived populations had appeared in China, as represented by fossils like those from Jinniushan and Dali. They seem more similar to African and European samples assigned to *Homo heidelbergensis* than to their local antecedents, so they may indicate the opening up of greater population movements or gene flow between the western and eastern parts of Eurasia during the later middle Pleistocene.

However, by this time in Europe, there were already clear signs of the emergence of early Neanderthals, since fossils from sites like Swanscombe, Ehringsdorf and, particularly, Atapuerca Sima de los Huesos show derived Neanderthal features in the cranial vault or, where preserved, the face. The Swanscombe material could be as ancient as 400 ka, while the Atapuerca material probably dates from around 250 ka, and Ehringsdorf from about 220 ka. The remarkably preserved sample from the Sima de los Huesos shows that the ancestral features of *heidelbergensis* were being lost, and the derived features of Neanderthals were being gained, in a gradual and mosaic fashion (e.g. reference 17). By about 125 ka, the whole suite of Neanderthal characteristics can be recognised in fossils from France, Italy and Croatia. These features include the voluminous and projecting nasal apparatus, coupled with laterally receding cheek bones, a large endocranial volume, particularly in actual or inferred male individuals, a subspherical skull shape when viewed from behind, and anatomical details of the occipital and temporal bones. Most recently, CT analyses have demonstrated that even the Neanderthal inner ear showed previously unsuspected distinctive features compared with other fossil and recent humans, and these distinctions must have been developing before birth, and therefore under genetic control.[12] In their postcrania, the Neanderthals retained the robusticity of the skeleton found in earlier humans, particularly in their legs, while their body shape apparently reflected the effects of the predominantly cold environments in which they lived. They had large and wide trunks, and relatively short extremities, physiques which are found in modern peoples from high latitudes, such as the Inuit and Lapps, and which minimise the surface area of the body from which heat could be lost.[18]

About 35 ka, new people appear in Europe, associated with novel stone tool industries and behaviours, such as body adornment (inferred from the presence of pigments, pendants and jewellery), symbolic art and complex burials. These people of the Upper Palaeolithic are commonly known as Cro-Magnons, after the French discoveries of 1868. Anatomically they were unquestionably much more like living people than were the Neanderthals—their cranial vaults were relatively higher and rounder in lateral profile, brow ridges smaller, faces smaller and flatter, and chins more prominent. Their skeletons, while still strongly built compared with the modern average, were reduced in robusticity compared with the Neanderthals, especially in their lower limbs. Their body proportions were more similar to those of warm-adapted people of today, despite the fact that they appeared in Europe in the middle of the last glaciation.[19] All of these data accentuate contrasts with the contemporaneous last Neanderthals, and one of the most intriguing questions still to be answered is what happened to them? Evidence suggests that they did not disappear synchronously with the appearance of the Cro-Magnons, but persisted in some parts of western Europe, at least, for a minimum of 5000 years. It therefore seems very unlikely that the extreme viewpoints which argue that Neanderthals either rapidly transformed themselves into modern humans, or were rapidly exterminated by the Cro-Magnons, are feasible. How, then, did these populations interact with each other? Over a time-scale of several thousand years, and over a landmass as large as Europe, it is possible to imagine many different scenarios, and indeed each of them may have occurred at some time or another. There may have been conflict between the groups, they may have avoided each other, they may have coexisted relatively peacefully and they may have had contact, such as through trade or even interbreeding. Overall, however, the Neanderthals appear to have been gradually marginalised, until by about 27 ka, they were completely extinct. A combination of economic competition from the Cro-Magnons for local resources and a series of rapid and extreme oscillations recorded in the climatic records of this period may have been the primary factors behind their demise.

2.3   THE AUSTRALASIAN RECORD

Thus it appears that modern humans did not originate in Europe—instead, the region shows a long sequence of regional continuity culminating in the Neanderthals of the last glaciation, followed by an intrusion of modern humans, and a relatively sudden replacement event. Let us now consider events over a comparable time-scale on the other side of the inhabited Pleistocene world, in Australasia. Even at the times of lowest sea level during recent glacial stages there has never been a land bridge connecting Australia with the islands of Indonesia, although New Guinea and Tasmania *were* part of an enlarged Australian continent at such times. Humans had colonised the islands of Indonesia by the early Pleistocene (1.5 ma), but seem not to have reached Australia until the late Pleistocene, less than 70 ka. The first colonisers were representatives of *Homo erectus*, and recent research suggests that they were in Java by 1.6–1.8 ma,[20] having presumably spread from Africa (a minority view is that the species actually originated from a more primitive local ancestor similar in form to African Pliocene fossils attributed to *Homo habilis*). There is then good evidence that *Homo erectus* persisted through the lower and middle Pleistocene in Java without very significant evolutionary change. Even more remarkably, recent dating work suggests that *Homo erectus* might still have been living in Java as recently as 30–50 ka from fossil evidence at the sites of Ngandong and Sambungmachan, on the banks of the River Solo. If these dates on associated animal teeth are accurate, then this ancient species survived in the south east of the inhabited world as long as Neanderthals did in the north west.[21,22]

People could only have arrived in Australia on water craft, and this would have entailed repeated island-hopping journeys of at least 50 km over open seas. The first Australians were no doubt unwilling colonists, carried by unkind winds or seas off course from an island they wished to visit to one they had never seen before. The exact route taken by these first colonists is unknown. There were at least two feasible routes—an eastern one via Timor to New Guinea, or a western one via Java to Northwest Australia itself. Archaeological sites in New Guinea have been dated to about 30 ka, but more recently, human occupation of inland rock shelters in northern Australia, Malakunanja II and Nawalabila, has been dated to at least 50 ka. More recently still, there have been claims that artefacts and rock art at a rock shelter called Jinmium date from at least 70 ka, and possibly much older.[23] However, the Jinmium results have not yet been verified, and other evidence suggests that the site was only occupied after 20 ka.

We have no evidence yet of the physical appearance of the first human occupants of Australia, since the earliest known fossils come from sites in the south east of the continent, in the Willandra Lakes region, and apparently date from a later stage of colonisation.[24] The large sample from there contains a cremated individual and a partial skeleton, both lightly built and possibly female, dated to at least 30 ka. However, there are also much larger and more robust individuals, and this has led to the creation of different scenarios to account for this marked physical variation. The simplest hypothesis is that only one founder population reached Australia; it then dispersed across a continent empty of people, and in doing so began to develop the physical variation which is evident at sites dating between 30–10 ka.[25] An alternative and more complex scenario suggests that there were two distinct founding populations. The more robust one colonised from western Indonesia, derived from late *Homo erectus* people, such as those known from Ngandong. The other more gracile arrivals came on the eastern route via New Guinea, and ultimately derived from Chinese *Homo erectus* ancestors.[1,2] It would be helpful in testing these hypotheses to know the age of the respective components of the Australian fossil record, but many important specimens are currently undated, and it is therefore impossible to ascertain the degree and direction of changes in robusticity. Moreover, there are problems with the interpretation of the fossil material itself because it appears that some of the most robust specimens have been modified by intended or unintended artificial deformation during life.[24] This has accentuated their flat frontal bones, increasing resemblances to their claimed ancestors, the late *erectus* people of Java. More difficult still, many of the most critical specimens in the debate about human origins in Australia have now been returned to aboriginal custodians for reburial, thus making them unavailable for further study. Progress can be expected over the chronology of the human occupation of Australia, however, through the further application of luminescence dating techniques to the sediments of the most ancient archaeological sites, and the use of non-destructive gamma ray dating on the remaining fossil record.

2.4   COMPARING THE TWO RECORDS: EUROPE AND AUSTRALASIA

Historically, the records of European and Australasian prehistory have developed in very different ways. The European record was subjected to intense study and debate almost from its inception, but even after the remarkable pioneering work of Eugene Dubois at the end of the last century, Australasian prehistory, like that of Africa, was slow to realise its potential. However, the time span of human history in Australasia can now be seen to greatly exceed that known for Europe, while even the arrival of modern humans in Europe appears to have been a relatively late event compared with the first colonisation of remote Australia. While the arrival of modern people in Europe seems to coincide with the first manifestations of Upper Palaeolithic technology, the archaeological record of Australasia appears not to show comparable developments until the last 10 000 years. Yet, other manifestations of 'modern' behaviour such as the use of boats or rafts, cremation of the dead, the production of art and body decoration were certainly apparent early on in Australian prehistory. Both Europe and Australia were probably colonised by small bands of early modern pioneers, but while the continent of Europe was previously inhabited (by the Neanderthals), that of Australia appears, in my opinion, never to have been reached by *Homo erectus*. On the other hand, recent evidence suggests that there might indeed have been contact between late archaic and early modern people in both regions, given that one modern human dispersal route into Australia could have been via Java, where it seems that *erectus* persisted into the relevant time frame. In both cases, the replacement of the local late archaic people followed, although details of the process in Java are lacking, since the first modern human remains, from Wadjak, appear to date from less than 10 ka.[24]

Morphological data suggest that early modern fossils from Europe and Australia share a number of features which may be primitive retentions,[26] including a relatively broad, flat face, with low orbits and nose, a relatively long cranial vault and (probably) a warm-adapted physique. They also shared many evolutionary novelties in cranial shape and in postcranial gracility compared with their local archaic predecessors, as well as lacking the distinctive derived features which the more ancient lineages in Europe and Java had developed. The most economical explanation of the shared evolutionary novelties is that they were inherited from a more ancient common ancestor, and a growing body of evidence points to Africa as the ultimate source of the modern human diaspora. However, the earlier dispersal of modern humans to Australia, as indicated by the available chronological data, would suggest a more complex dispersal pattern from Africa, one of at least two phases.[27] The first phase, perhaps reflected by the presence of modern humans in the Levant at about 100 ka, consisted of the eastward movement of people through subtropical regions into South-east Asia, and was achieved with Middle Palaeolithic technology. With the invention of water craft, this was subsequently sufficient to take them all the way to Australia. Surviving signs of the anatomy of these pioneers can probably still be recognised in common features of dental morphology shared between present day native Australians and sub-Saharan Africans.[28] A subsequent Eurasian dispersal of early modern humans who had lost some of those dental features, and had acquired an early Upper Palaeolithic technology, took place around 40 ka.

2.5 GENETIC DATA

The growing body of genetic data certainly predominantly supports the model of a recent African origin for modern humans (e.g. reference 29), although there are also indications of more complex patterns, such as possible gene flow between archaic and early modern humans in Asia, and movement of genes or people into, as well as out of, Africa.[30] It has become evident that many phylogenetic reconstructions from genetic data are also dependent on assumptions about past demographic history, such as ancient population sizes.[31] It has also become clear that effective human population size has probably fluctuated dramatically during the Pleistocene,[32] and probably suffered a dramatic decline before modern humans finally emerged from Africa. Perhaps the environmental effects of the long cold period between 200–130 ka severed any remaining genetic connections between archaic human populations in Eurasia and Africa, leading to the final isolated development of late Neanderthal and early modern human populations in their respective European and African homelands.

Testing such hypotheses from genetic data always seemed an impossibility if analysable DNA came only from the extant survivors of such events—living *Homo sapiens*. Now, with the recovery of DNA from the type specimen of the Neanderthal species, we can start to go beyond extrapolations from the present and look directly into the past.[33,34] The mitochondrial DNA pattern of the Neanderthal does indeed show that human mtDNA diversity was much greater in the past, and allows a calibration of the divergence time of the Neanderthal pattern from that characterising modern humans of about 600 ka. Gene divergence precedes population and species divergences, but this figure is certainly compatible with interpretations from the fossil record that the Neanderthal lineage separated from our own at about 300 ka. Equally, it is incompatible with suggestions that Neanderthals

were either uniquely ancestral to recent Europeans through evolution, or were partly ancestral through hybridisation. Of course,
it is to be hoped that many more such sequences will now be
obtained, not only from other Neanderthals and archaic peoples,
but also from early modern samples in various parts of the world.
European Cro-Magnon fossils are already being tested, and it
would be a tremendous achievement if the equally ancient and
fascinating human fossils of Australia could be made to yield up
whatever DNA secrets they still preserve. Let us hope that the
recent astonishing level of breakthroughs in discoveries, dating
and DNA detection can be maintained, and that in this case we
will be able to view the old Chinese curse 'May you live in interesting times' as more of a promise than a threat!

3

Language classification: scientific and unscientific methods[1]

Don Ringe

What do we linguists have to say—or what should we have to say—about our common human inheritance? I will argue that we have less to contribute than is often supposed; but we can make some contribution, because of the way languages are usually classified. We classify languages into families, not on the basis of general similarities between them, but on the basis of shared history. Some of those historical connections are documented, but many more can be inferred to have existed in the prehistoric period. Because a family relationship always implies a historical connection, and because so many of those connections are in fact prehistoric, it is only natural to try to use the classification of languages into families to draw inferences about human prehistory, and people have been doing that for well over a century. Unfortunately it has sometimes been done with more enthusiasm than judgement; not everyone seems aware that there are severe limits to the inferences that can be drawn about prehistory from language relationships. In particular, Joseph Greenberg, Merritt Ruhlen, and other proponents of 'long-range' comparison have made strong claims about language relationships and classification in the past decade or so, and those claims have promptly been made the basis of inferences about human prehistory. That is unfortunate, because the evidence simply does not support the claims in question. To understand why that is so, one has to understand how language families can be established objectively; therefore this paper will discuss the standard, well-established method by which languages are classified, as well as the alternative proposal favoured by 'long-rangers'.

A language family is, by definition, a group of languages that were once a single language.[2] The standard examples from recorded history are well known: Latin developed into the Romance languages; Classical Arabic developed into the modern Arabic languages (or 'dialects'; but they are very diverse dialects, some of which can easily be described as separate languages[3]); Middle Chinese developed into the modern Han languages. (I use these families as examples because their diversification can be followed in the historical record to a considerable extent. All three are of course subfamilies of still larger families.[4]) This is not the only type of historical connection between languages that can exist, much less the only reason why they can resemble one another;[5] but for the subject of this book it is the most important type of language relationship, and it is the usual basis of language classifications.

Even from such a cursory description it is clear that language families have a great deal to do with language change; the reason why there *is* a Romance family, for example, is that Latin changed, from generation to generation, one set of ways in central Italy, another in northern France, a third in Castile, and so on. Thus any discussion of language classification is also a discussion of language change in past centuries; and any discussion of language change, or of any other process or structure in human language, must be grounded in a general principle of linguistics: *all scientific linguistics is based on observation; that is, all hypotheses must be grounded in observation and must be accountable to all relevant facts.* Of course linguistics does involve inferences, as all systematic studies based on observation must, but properly rigorous linguistics does not involve speculation; all inferences must be based on observational data. If one rejects the objective and observational nature of linguistic enquiry, or does not deal with the linguistic facts in detail, one has nothing to contribute to a discussion of the structure or history of human language, or of language classification—just as anyone who rejects an objective basis for biological study, or who has not seriously studied biology, has no basis for an opinion about whether the pattern of speciation is one of punctuated equilibria, or about any other important biological question.

On the other hand, if we take the objectivity of linguistics seriously, we have to ask the hard question: how on earth can we bring observation to bear on what was happening in Paris in the dark ages, or what happened in stone-age Europe? There is only one way to do it, and that is strict adherence to the *uniformitarian principle*: unless we can demonstrate that conditions of language use have been altered in such a way as to affect language structure and change, we must posit for unobservable language communities the same types of structures and changes that we observe in the historical record and at the present time.[6]

In other words, we use the same principle that geologists and palaeontologists use to bridge gaps in our historical record and to extrapolate into prehistory. In fact we may be able to make better use of the uniformitarian principle than they can. The pace of geological and evolutionary change is very slow, relative to human lifetimes, and one has to reckon with the possibility of events (such as ten-kilometre bolide impacts) so rare that the whole of human history provides not a single example. By contrast, language change is perceptible even in a series of documents that span a few centuries, and some changes spread through a speech community fast enough that there is an easily noticeable difference between the speech of the oldest and youngest members of the community (as the work of Bill Labov has repeatedly shown). So we ought to have, somewhere in our records, examples of almost everything that can happen in language change—and far more than enough material to make valid statistical arguments about what is likely or unlikely to have happened in prehistory.

Notice what this implies: everything in linguistics is relevant to our hypotheses about prehistory. A hypothesis that is clearly incompatible with anything that is already certainly known must be rejected. That is one of the major reasons why proposals about linguistic prehistory from outside the field have been received with so little enthusiasm by linguists: whatever the individual arguments may sound like, it always comes down to a clear judgement by a large majority of linguists that the hypothesis is seriously incompatible with something that is certainly known—and thus violates the uniformitarian principle. Such judgements are not intended to be exclusionary; they are necessary to preserve the integrity of the field, because if we abandon the uniformitarian principle we have no basis at all for scientific historical linguistics.

But how much is actually known about language structure and change? I will offer two answers. One is to ask the reader to examine the standard bibliography of linguistics, the *Bibliographie linguistique*. For each of the last three years in which it appeared, 1992 through 1994, it listed more than 20 000 titles; it has been listing more than 12 000 titles per year since 1970, and even the first volume in which titles were numbered (1962) listed more than 9 000. Not that everything listed is good work, of course; but an enormous amount of work has been done and an enormous amount has been learned, especially in the last generation. In fact, if one's information is more than a decade old, one is certainly out of date, because the field is still developing very fast. And as we learn more and more about how human language works, how children acquire language, what happens in bilingual communities, and so forth, our standards necessarily become more and more stringent; sometimes a hypothesis that seemed at least possible twenty years ago can be shown to be utterly unrealistic in the light of modern research.

The other answer to the question 'How much is known?' will be the rest of this paper, in which I will examine two methods of classifying languages, one based on an exact understanding of the processes of language change, the other apparently based on common sense.

First I need to outline what is known about language *change*. So far as we can determine, all living languages change all the time. A reasonable reaction to that fact is surprise; after all, a system of communication would seem to be much more useful if it were stable. But constant change is what we observe, both in the historical record and in contemporary speech communities. (One of the most startling findings of modern work in sociolinguistics is that neither writing nor the mass media have had any significant effect on the occurrence of language change; it continues to occur, rapidly and vigorously.) Moreover, change affects all aspects of a language's structure—and language is mostly structure; 'words' may be the elements of a language that are easiest to grasp, but they are actually the least significant parts structurally, as can be seen from the fact that languages borrow individual words from one another all the time without any alteration in their overall structure. Thirdly, linguistic changes are demographically specific; that is, they typically occur in a single dialect of a single language spoken by a single social group in a single community during a restricted span of time (a few generations). Of course changes can and do spread, through contact between social groups and communities, but the point is that they don't have to; in principle, every speech community shows its own distinctive pattern of historical changes. This immediately explains the existence of language families: if a single speech community splits into two or more—either because of migration, or because the community has become too large (geographically) for all the subcommunities to stay in effective touch with one another, or because the community is sharply divided along social lines—and if the new communities lose contact, even partially, each will undergo idiosyncratic changes of its own.[7] Over time the changes add up; if contact remains minimal or nonexistent, so that the changes do not spread from group to group, eventually the forms of speech in question will become so different that they have to be called separate languages, and a single language will have developed into a family. The analogies with biological speciation are obvious, but linguistic speciation happens far more quickly.

But there is one more thing to be said about language change, and it is the biggest surprise of all: unlike other linguistic changes, changes in pronunciation—'sound changes', to use the technical term—are almost entirely regular. In any given line of linguistic descent[8] in a given period of time, one of two things will happen to each distinctive sound: either it will develop the same way in every word in which it occurs; or, if there are factors that condition particular developments, they can be stated entirely in terms of other sounds in the same word or phrase.

This can be illustrated by a fairly simple example from the history of English. Old English (OE; i.e., Anglo-Saxon, the stage of

English attested before the Norman Conquest) had a long vowel /ā/ (roughly as in modern English *father*).[9] In the southern dialects of Middle English this vowel was rounded to /ɔ/ (roughly as in modern *broad*); then, in the 'Great Vowel Shift' of the late 15th century it was raised to /ō/, and eventually it was diphthongized to /ow/, the modern sound.[10] Every one of these changes affected all the relevant words in the language, unless specific phonetic factors intervened; consequently the confrontation of OE and Modern English reveals a regular pattern:[11]

sāpe > soap	bān > bone
hlāf 'bread' > loaf	hāl 'healthy' > whole[12]
hām > home	āc > oak
gāt > goat	dāg > dough
rād 'journey' > road	fāh 'hostile' > foe
āþ > oath	snāw > snow
gāst 'spirit' > ghost	

There are dozens of examples that show this pattern. A similar regularity obtains for OE /ū/ (roughly as in modern *food*), which survived unchanged through most of the Middle Ages, was diphthongized to /əw/ in the 15th century, and ultimately became modern /aw/:

hūs > house	ūre > our
ūt > out	fūl > foul
brūn > brown	nū > now
hlūd > loud	mūþ > mouth

Again, examples can be multiplied. But these sets of regular sound changes were not completely unconditioned; under some phonetic conditions, which are precisely specifiable, other changes intervened. For instance, in early Middle English both these vowels were shortened in the antepenultimate syllables of words, and before most clusters of two or more consonants (though not before /st/, for example); the resulting short /ɔ/ later became (American) /a/, and the resulting short /u/ became /ʌ/. Note the following examples:

holiday /halədey/, but holy /howli/ < OE hālig
bonfire /banfajər/, but bone /bown/ < OE bān
southern /sʌðərn/ (ME sutherne, three syllables), but south
　/sawθ/ < OE sūþ
husband /hʌzbənd/, but house /haws/ < OE hūs
utmost /ʌtmowst/, but out /awt/ < OE ūt

Further, the diphthongization of /ū/ failed to occur when the
vowel was immediately followed by a labial consonant (i.e., a consonant produced with the aid of one or both lips); the surviving
/ū/ was eventually shortened to /u/, which became /ʌ/ (as
above). Note the following examples:

plum /plʌm/ < OE plūme	dove /dʌv/ < OE dūfe
thumb /θʌm/ < OE þūma	shove /šʌv/ < OE scūfan
sup /sʌp/ < OE sūpan	up /ʌp/ < OE ūp

In short, though OE vowels have changed their pronunciation
a good deal in the past millennium, they have done so in regular
patterns, 'regularity' being defined exclusively in terms of sounds.

The regularities just exemplified are not quite absolute, but they
come startlingly close to perfection. A few hours' work with dic-
tionaries enabled me to estimate that, of the OE words that survive
in modern English, about 94% exhibit completely regular development of all their sounds. Moreover, in those words that exhibit
some irregularity, it typically affects the development of a single
sound; and since the average number of sounds per word is about
four (because it is mostly the shorter OE words that survive), the
regularity of development in terms of tokens of sounds in the
lexicon is greater than 98%.

This very high degree of regularity in the development of
sounds is typical; it is what we find in every language whose history we can trace at all. The only thing strange about this is that I
was able to illustrate the point even with standard English examples. Recall how the regularity of sound changes operates: we
expect to find regularity within each line of linguistic descent,
defined roughly as a speech community. But we have plenty of
evidence that standard Modern English does *not* represent a single

line of linguistic descent—a single speech community; on the contrary, it reflects massive interaction between the dialects spoken by several different communities whose members gravitated to London beginning in (probably) the 14th century. In other words, I have taken my examples of regular sound change from an outstandingly unpromising source, in which substantial *ir*regularity should be evident (and there certainly is some). In languages which have been less affected by dialect interactions, the observed regularity of sound changes is even greater; in some it is almost total. In asserting that the regularity of sound change is overwhelming, I am not idealizing, nor making a programmatic statement; this is a straightforward observation about facts that have been known to linguists for more than a century.

Readers who have some knowledge of intellectual trends in historical linguistics may be under the impression that the regularity of sound change has been questioned. Whether that is true depends very much on what one means. The overwhelming statistical regularity of sound change has never been in doubt, any more than the reality of biological evolution has been doubted by modern mainstream biologists. As usual, the argument is more theoretical: whether the regularity of sound change can be elevated to a general principle, how it works out in the real world of speaking populations, and why sound change exhibits such overwhelming regularity are all questions that have been vigorously debated. In every generation since about 1870, close investigation has at least vindicated the regularity of sound change as a phenomenon that needs to be explained. Probably the most up-to-date discussion of all these questions is Labov 1994, to which readers are referred. For practical work in historical linguistics, though, the simple statistical fact of overwhelming regularity is by far the most important thing about sound change, because of the way we can exploit it to recover information about linguistic prehistory. Here is how we do that.

Given that sound changes are overwhelmingly regular in each line of development, it follows that there will be regular *correspondences* between the sounds of languages within a single family— to the extent that they have not replaced the words and forms that they inherited from their common ancestor with completely different words and forms (something which all languages also do). A diagram of part of the Romance family may help to make this clearer:

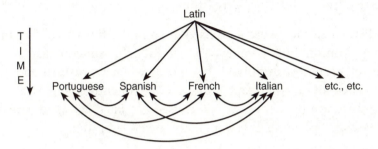

Since the development along each of the single-headed arrows—that is, from Latin to each of the Romance languages—proceeds by regular, mechanical changes in the sounds of words and forms, and since each line of development starts from the same system of words and forms, there will inevitably be regular correspondences between the sounds in the inherited words and forms of any pair of Romance languages—that is, along each of the double-headed arrows in the diagram.

Here are a few examples from French and Spanish. Usually Spanish *e* corresponds to French *oi* /wa/,[13] because Latin stressed *ē*[14] became *e* in Spanish but *oi* in French in most phonetic environments:

Latin	French	Spanish
dīrēctum[15] 'straight'	droit 'straight, right'	derecho (both mngs.)
pēnsum 'weighed'	poids[16] 'weight'	peso 'weight'
stēllam 'star'	étoile	estrella
tēctum 'roof'	toit	techo
tēlam 'web, warp'	toile 'linen'	tela 'cloth'
trēs 'three'	trois	tres
vēla (pl.) 'sails'	voile (sg.) 'sail'	vela (sg.)

Similarly, Spanish *ie* /je/ usually corresponds to French *ie* /jɛ/, because Latin stressed (short) *e* underwent the same sort of diphthongization in both those languages:

Latin	French	Spanish
fel 'gall bladder'	fiel 'bile'	hiel 'bile'
herī 'yesterday'	hier	ayer (a 'at, to')
mel 'honey'	miel	miel
petram 'rock, cliff'	pierre 'stone'	piedra 'stone'

But when they were *un*stressed in Latin, both e-vowels became *e* in Spanish, but '*e* muet' /ə/ in French; subsequently those vowels were lost in polysyllabic words when word-final and not preceded by a consonant cluster. Consequently there is also a regular correspondence 'Sp. unstressed *e* = Fr. *e* /ə/' (and, strictly speaking, 'Sp. word-final Ø (i.e., no vowel) = Fr. Ø'):

Latin	French	Spanish
dēbēˊre 'to owe'	devoir_ 'to owe; ought'	deber_ (both mngs.)
veníre 'to come'	venir_	venir_
légem 'law'	loi_	ley_
régem 'king'	roi_	rey_
ménsem 'month'	mois_	mes_
léporem 'hare'	lièvre	liebre

(Note especially French stressed *moi,* unstressed *me,* both directly descended from Latin *m ē*'me'; the same stress conditioning accounts for the doublets *toi* and *te, soi* and *se.*) Further, while the diphthongization of stressed short *e* occurred 'across the board' in Spanish, it did not occur in French before certain Latin consonant clusters, notably *rC* and *sC;* in those environments modern French therefore has /ɛ/ rather than /jɛ/. Thus there is also a regular correspondence 'Sp. *ie* = Fr. /ɛ/' in the appropriate consonantal environments:

Latin	French	Spanish
ferrum 'iron'	fer	hierro
herbam 'grass'	herbe	hierba
terram 'land'	terre	tierra
testam, testum 'clay pot'	tête 'head'	tiesto 'flowerpot'

This by no means exhausts the developments of Latin e-vowels in French and Spanish, which were very complex; yet the result is a regular and easily discoverable pattern of sound correspondences between those languages.

Now consider a situation which we very frequently encounter in historical linguistics: we find such a pattern of regular correspondences between the sounds of a group of languages, but the historical records of those languages do not directly establish a connection between them. Here is a small part of such a pattern of correspondences from some well-known languages of northern Europe:

	English	Netherlandic	High German	Icelandic
(set 1)	bone	been 'leg, bone'	Bein 'leg'	bein 'bone'
	dough	deeg	Teig	deig
	ghost	geest 'spirit'	Geist 'spirit'	—
	home	heem 'farmyard'	Heim	heimur 'world'
	oak	eek	Eiche	eik
	oath	eed	Eid	eiður
	soap	zeep	Seife	—
	stone	steen	Stein	steinn
	whole	heel	heil 'unhurt, intact'	heill 'whole'
(set 2)	brown	bruin	braun	brúnn
	foul	vuil 'rotten'	faul 'lazy'	fúll 'foul'
	house	huis	Haus	hús
	loud	luid	laut	—
	mouse	muis	Maus	mús
	out	uit	aus	út
	town	tuin 'garden'	Zaun 'hedge'	tún 'farmstead'

As the reader can see, I have compared some English words containing the phonemes /ow/ and /aw/ with words of the same or similar meaning in some other languages; since those languages have consistent orthographies (in which the same phoneme is normally spelled the same way), it will be possible to compare the vowels without further analysis. We have two regular vowel correspondences: English /ow/ corresponds to Netherlandic *ee*,

German *ei,* and Icelandic *ei,* while English /aw/ corresponds to Netherlandic *ui,* German *au,* and Icelandic *ú.* (Of course this is only a tiny part of the pattern; vowel correspondences between these languages are just as complex as those between Romance languages.) The wordlists would resemble one another much more if I had used mediaeval rather than modern stages of the languages, but even so they would not be identical; and of course all these languages emerge into written history as separate languages. Yet they show the same sort of regular sound correspondences as the Romance languages do. By the uniformitarian principle we must suppose that the pattern arose in the same way: that is, that these languages (and some others) form a family—the Germanic family—which was once a single language—Proto-Germanic—even though we have no records of that 'parent' language (as we do of Latin) because the speakers of Proto-Germanic were illiterate barbarians.

In other words, we exploit the overwhelming regularity of sound change and the uniformitarian principle to recognize language families in groups of languages showing regular sound correspondences. We can do more; in particular, we can use a simple mathematical method to triangulate backwards in cases like this, undoing the sound changes and reconstructing part of the unattested parent language. But discussion of that procedure is beyond the scope of this paper; here I must concentrate on language classification.

All scientific language classification is based on the methodology outlined above. No one denies that this is the surest way to investigate language prehistory and classify languages (though of course we do encounter instances in which the evidence is inadequate, so that this method cannot be effectively applied). So far as I know, even proponents of long-range comparison accept this. For example, in a recent book that offers a controversial classification of the languages of the Americas (Greenberg 1987),[17] Joseph Greenberg accepts without discussion the families of American languages that were established by the methodology just described. However, he expresses consternation about the (admittedly large) amount of work that the standard methodology involves,[18] and he claims to have discovered a useful 'short cut' which is based not on the detailed study of language structure and change, but apparently on common sense. In order to evaluate that claim, and similar claims made by other proponents of long-range classifications, we must first examine the relationship between the rigorous findings of scientific historical linguistics and common sense.

First of all, even though we can show that sound changes are regular, it is reasonable to ask whether we need to go to so much trouble to find families of languages. In all the examples we have seen so far, the similarities between the related words have been moderately obvious. Can't we simply work with those? Before we attempt to answer that question, let us consider carefully what it presupposes. Such a strategy will work only if sound changes are not only regular but also moderate, effecting relatively small changes in pronunciation, and do not occur too often, so that the effects of many sound changes do not accumulate so quickly as to disguise cognates within a millennium or two; only under those conditions could a basic knowledge of phonetics allow us to iden- tify sets of related languages simply by inspection of comparative data, without working out the regular sound changes.

Unfortunately we cannot count on encountering such 'linguist- friendly' conditions; in particular, it sometimes happens that over relatively short periods of time the effects of sound changes accu- mulate to cause substantial differences in the sounds of provably related languages. Spectacular examples of this are provided in abundance by Arapaho, a language spoken in northeastern Colorado since before the time of European contact. The only obvious relative of Arapaho is Atsina (also called Gros Ventre), a neighbouring language so similar that they might almost be called dialects of a single language; but further investigation eventually revealed that Arapaho and Atsina are members of the Algonkian family, most of whose languages were spoken further east (around the Great Lakes, in eastern Canada, in New England and the areas immediately to the south). What made the membership of these two languages in the Algonkian family so hard to see was the fact that they have undergone radical sound changes in the last few centuries, as a few Arapaho examples will demonstrate. For

instance, in a wide range of phonetic environments the *p inherited from Proto-Algonkian has become č (as at the beginning and
end of English *church*)—a very different sound—giving rise to a
regular sound correspondence that is very unusual phonetically:

Arapaho	**other Algonkian languages**[19]	
čeʔí-θe: 'ashes' =	Ojibwa pinkwi	< PA *penkwi
čebíte: 'grease' =	Fox pemite:wi, Oj. pimite:	< PA *pemite:wi
čê:séy 'one' =	Oj. pe:šik	< PA *pe:šekwi
téče ʔ 'night' =	Fox tepehkwi[20]	< PA *tepexkwi
néč 'water' =	Fox nepi, Oj. nimpi	< PA *nepyi
ni:čí: 'river' =	Fox si:po:wi	< PA *si:po:wi
néíčiθ 'my tooth' ≈[21]	Fox ni:piči	< PA *ni:piči

Under the same conditions inherited *m has become *b* (which
is not so drastic, though still a fairly unusual change); compare the
following examples:

Arapaho	**other Algonkian languages**	
béʔ 'blood' =	Fox meškwi, Oj. miskwi	< PA *meçkwi
ben- 'drink'	cf. Fox menowa '(s)he's drinking'	< PA *men-
bi:θí-hiʔ '(s)he's eating',	cf. Fox mi:čiwa, Oj. mi:čin	< PA *mi:či-
bí:í 'feather' =	Fox mi:kona	< PA *mi:kona
bi:n- 'give' =	Fox, Oj. mi:n-, Unami mi:l-	< PA *mi:l-
beʔe:- 'be red' =	Fox meškwa:-, Oj. miskwa:-	< PA *meçkwa:-
čebíte: 'grease' =	Fox pemite:wi, Oj. pimite:	< PA *pemite:wi

Some of the sound changes that occurred in Arapaho are even
stranger. Readers who examine the above lists in greater detail will
find, for example, that PA *č, preserved in most Algonkian languages, has become θ (as at the beginning of English *think*) in
Arapaho; that PA *k has vanished without a trace (!; cf. 'one',
'feather'), but that any immediately preceding consonant survives
as the glottal stop ʔ (which is like the Liverpudlian pronunciation
of *tt* in *bottle;* cf. 'ashes', 'night', 'blood', 'be red'); short *o has
merged with *i ('feather') and long *o: with *i: ('river'); and so

on. Interested readers should consult Goddard 1974 for fuller information.

Moreover, though Arapaho is not a run-of-the-mill case, it is not unique; in our own Indo-European family we have at least one language notorious for its bizarre sound changes, namely Armenian (cf. Hübschmann 1897, Godel 1975, Schmitt 1981). The Armenian facts are well known; I mention only a couple of the more outstanding oddities here. Armenian k^h- corresponds to Sanskrit sv- and Greek h-, all three having developed from an original Proto-Indo-European consonant cluster *sw-, though the stages of the Armenian development are difficult to reconstruct:

Arm. k^hown 'sleep' = Skt. svápnas, Gk. húpnos < PIE *swépnos
Arm. k^hoyr 'sister' = Skt. svásā < PIE *swésōr
Arm. k^hirtn 'sweat' ≈ Gk. hidrŏ́s, Skt. (verb) svid- < PIE (verb) *swid-, (noun) *swid-ró- (extended by further suffixes in Arm. and Gk.; note that the order of r and t < *d has been reversed regularly (!) in Arm.)

Still more surprisingly, Armenian erk- corresponds to Skt. dv-, Gk. d-, reflecting Proto-Indo-European *dw-:

Arm. erku 'two' = Skt. dvắ < PIE *dwŏ́ (cf. Gk. dŏ́deka 'twelve')
Arm. erkar 'long' (adv.) = Gk. dērón < PIE *dwāróm
Arm. erki- 'fear' = Gk. di- 'fear', Skt. dvi-ṣ- 'hate' < PIE *dwi- 'fear'

Eccentric cases of this sort occur often enough that if we try to find the provable language families merely by 'eyeballing word-lists', we are going to miss a significant number of relationships that are perfectly clear and uncontroversial.

But what if we turn the question the other way round? If we find clear similarities in form and meaning between some basic words and inflections of two languages, can we take that as an indication of relationship without working out the sound changes? Again the answer is no, this time for a very simple reason: similarities between languages arise by sheer chance startlingly often; to the casual observer, using only common sense as a guide, they can seem preposterously common. Here are some examples from

English and Spanish—two languages whose histories and prehistories are so thoroughly researched that we can say with complete confidence whether or not there is any connection between their words.

English-speaking learners of Spanish commonly suppose that *mucho* is etymologically equivalent to English *much;* in fact there is no connection at all between them, and the historical record demonstrates that without any doubt whatever. Here are the histories of the two words (in reverse) for the last two thousand years:

ModE much < (southern) ME muchel 'big' (cf. the shortening of ME wenchel to ModE wench) < OE micel < Proto-Germanic *mikilaz (cf. Gothic mikils); the word is ultimately related to Gk. mégas and Lat. magnus, which likewise mean 'big'.

Sp. mucho < *muyto (cf. Portuguese muito) < Lat. multum (parallel to Sp. escuchar 'to listen' < *scuytar < Lat. auscultāre); the word is ultimately related to Gk. mála 'very' and Lat. melior 'better', but there are no Germanic cognates.

Even if one knows nothing at all about linguistics, a basic pattern in these developments reveals that the words can't be related: as we trace them backward in time, they appear less and less similar—not at all what one would expect if they shared an origin in some particular prehistoric word; their present similarity is patently the result of convergent evolution. We are looking at what a biologist would call analogous rather than homologous traits. It is easy to find other examples from the same languages that show the same telltale pattern of convergence:

(a) ModE iron /ajərn/ < OE īren < īsern < PGmc. īsarnã (cf. Old High German īsarn); the word occurs only in Gmc. and Celtic.

Sp. hierro /jeřo/ 'iron' < Lat. ferrum (see the Romance cognate sets above); the word occurs only in Latin and its descendants.

(Cf. Sp. h (now silent) < Lat. f also e.g. in hijo 'son' < fīlium, etc.; the odd OE consonant change also in ūre 'our' < ūser < PGmc. *unseraz, cf. OHG unsar.)

(b) ModE fire < OE fȳr < PGmc. *fuir (cf. OHG fuir), ultimately
related to Gk. pûr (no Lat. cognate).

Sp. fuego < Lat. focum 'hearth', found only in Lat. and its
descendants. (Lat. f survives in Sp. before ue, cf. e.g. Sp. fuente
'fountain' < fontem; Engl. f normally corresponds to Sp. p, cf.
e.g. ModE father = Sp. padre, ModE fish = Sp. pez, etc.)

(c) ModE day < OE dæg < PGmc. *dagaz (cf. Goth. dags, ON
dagr); the word is found only in Gmc.

Sp. día < Lat. diēs (with remodelled ending); cognate with
Old Church Slavonic dĭnĭ, etc.—all the cognates show di-
with various suffixes, and none contains any g or similar consonant.

(Sp. initial d normally corresponds to Engl. t, cf. e.g. Sp. dos =
ModE two, Sp. diez = ModE ten, Sp. diente = ModE tooth,
etc.)

This sort of development is not rare; it happens literally all the
time. Every historical linguist specializing in some particular language family has a favourite set of such convergences with which
to amaze bored undergraduates.

Now what exactly is the commonsense shortcut of long-range
comparison? Unfortunately it is neither more nor less than superficial inspection of parallel wordlists to find words that are superficially similar in sounds and meaning—a procedure fatally subject
to the pitfalls just illustrated. For example, Greenberg claims that
whereas it is easy to find numerous illusory lookalikes between any
pair of languages, simultaneous examination of several dozen—or
several hundred—languages obviates the problem. Let us consider
this claim in detail. It is true that if we are examining even ten or
twenty parallel wordlists and we find, for a given meaning (say, the
basic concept 'hand'), very similar words in every single wordlist,
that can't be the result of chance—so far, so good. But what if we
are examining more than 900 wordlists, as Greenberg did in pre-
paring his 1987 book, and we find only ten or twenty lookalikes in
a given meaning among the whole set of 900-plus lists? Greenberg
has suggested that the odds of finding ten out of ten are the same
as finding ten out of 100, or 900, or any number (Greenberg

1963:3); that appears to be the basis of his claim that similarities between multiple languages are always greater-than-chance.

But that calculation of the odds does not work, as can be demonstrated with ease using standard probability theory. Let us suppose that, in a particular language, the probability of some particular consonant showing up at the beginning of a word chosen at random is .06, because that consonant appears word-initially in 6% of the words in the language—a very realistic premise. Let us suppose further that the same consonant appears word-initially in 6% of the words in each language whose parallel wordlists we are examining; that is less realistic in detail, but an average incidence of 6% is still very realistic for many consonants. What if we examine only five lists and find that, for a particular word, this consonant appears word-initially in every single language? The odds of that happening by sheer chance are $.06^5$, or less than one in a million—low enough[22] that, if we suggested a relationship between the five languages even on the basis of our single apparent cognate set, further investigation would most certainly be warranted. But suppose we are examining ten wordlists rather than five: what are the odds of getting a match in *any five* of the ten languages (or any four, or six, or some other specified number)? This is a well-known elementary problem, addressed in the first few chapters of any textbook on probability theory; I will spare you the calculations and give you the standard results:

no. of lgg. in the correspondence	% of words[23]	cumulative %
0	53.9	53.9
1	34.4	88.2
2	9.9	98.1
3	1.7	99.8

As can be seen, if we are comparing only ten wordlists Greenberg's claim is still reasonable: since the consonant appears word-initially either in no language or in only one in about 88% of the words in the list, we expect to find a chance correspondence—that is, the same initial consonant appearing in words of the same meaning in two or more languages *by chance*—for about 12% of the words in the list, and a chance correspondence involving *three or more* of the languages should appear in only about 2% of the words in the list; so it's clear that five-way correspondences will appear by chance very seldom. But if we increase the number of languages compared to only fifty, the situation is already very different:

no. of lgg. in the correspondence	% of words	cumulative %
0	4.5	4.5
1	14.5	19
2	22.6	41.6
3	23.1	64.7
4	17.3	82.1
5	10.2	92.2
6	4.9	97.1
7	2	99.1
8	0.7	99.7

In such a case we expect to find a chance correspondence between two or more languages for more than 80% of the words in the list, a chance correspondence between three or more languages for almost 60% of the words, a chance correspondence between four or more languages for more than a third of the words, between five or more languages for about 18% of the words—nearly word one in every five—and so on. If we really want to exclude chance similarities, we are going to have to demand correspondences involving at least seven of the 50 languages. And if we increase the number of languages compared to 100, the incidence of chance similarities increases dramatically:

no. of lgg. in the correspondence	% of words	cumulative %
0	0	0
1	0.1	0.1
2	0.4	0.5
3	1.3	1.9
4	3.1	5

5	5.8	10.8
6	9	19.8
7	11.8	31.7
8	13.5	45.2
9	13.6	58.7
10	12.2	71
11	9.9	80.9
12	7.3	88.3
13	5	93.2
14	3.1	96.3
15	1.8	98.1
16	1	99.1
17	0.5	99.6

Comparing 100 languages we expect to find chance correspondences between five or more languages for 95% of the words
in the list, and if we really want to exclude chance similarities, we
have to demand correspondences involving at least 15 or 16 of the
languages. Note that we are still working with only 100 languages,
whereas Greenberg worked with more than 900 in researching
Language in the Americas.

These simple and straightforward mathematical arguments are,
in my opinion, fatal to the enterprise of long-range comparison as
currently practised.[24] But of course each case must be judged individually on its merits, since it is always possible to happen upon a
genuine instance of linguistic relationship using unreliable methods; it might therefore be instructive to examine the results of
some particular long-range study. An attempt at long-range classification that has been widely discussed in the past decade is
Greenberg 1987, which claims that most of the 150 or so language
families of the Americas are actually subgroups of a vast 'Amerind'
family.[25] I examined the entire 'Amerind etymological dictionary'
included in Greenberg 1987 using the methods just outlined, and
my results (reported in detail in Ringe 1996) can be summarized
in a single sentence: the entire pattern of correspondences
reported in the 'Amerind' dictionary, on which the classification
of Greenberg 1987 is based, falls comfortably within the range of

chance resemblances. Only the approximately 150 well-established language families of which 'Amerind' is composed are demonstrable historical entities; the entire larger construct could be a mirage. This demonstrates the fallibility of long-range comparison most forcibly; even I was surprised at such a result.

Given that the shortcut of long-range comparison has gone so disastrously wrong, is it possible that more reliable shortcuts exist? It most certainly is, and it is perfectly obvious where to look for them: we should try to devise tests of whether particular similarities between languages are greater than could reasonably have arisen by chance even if they are not great enough to provide suitable material for a full application of the standard comparative method. Both Johanna Nichols and I have been working on that problem for several years. Nichols has interesting results already in print (see especially Nichols 1996). Briefly, she is testing complex and peculiar systems of grammatical affixes, attempting to provide realistic estimates of whether a particular configuration or system could reasonably have arisen by chance more than once; if the answer is *no,* the best hypothesis is that that configuration did arise only once, and that two or more languages or families that share it are historically connected, in some way or other, to the single language in which it arose. I have been attempting pairwise probabilistic analyses of comparative vocabularies. I originally thought that that would be fairly simple, but hard experience has taught me better. I take this opportunity to say that my initial publication on the subject (Ringe 1992) is now effectively obsolete; it has become obvious that it presents merely a crude approximation of the mathematics which cannot usefully be generalized, and the various corrections that I have made in subsequent publications (Ringe 1995 and forthcoming) only partly remedy the situation. I have therefore begun to rethink the problem, and I believe that work currently in progress has achieved a practical solution (though not all the theoretical problems have been solved). I hope to publish that work, which reaches the same results as my older work by means of more accurate mathematics, in the near future.

Of course such mathematical analyses cut both ways: it is very exciting when they suggest an unsuspected or unproven relationship between established language families, but they can also suggest that reasonably plausible hypotheses don't measure up. A case in point is the 'Nostratic' hypothesis, which posits a family relationship between the Indo-European, Uralic, and Kartvelian families, and some other families of Eurasian languages, at a very great time depth (certainly more than ten millennia in the past). Both Johanna Nichols and I have investigated parts of the Nostratic proposal, and both of us have concluded that those parts do not appear to be provably correct, simply because the linguistic similarities in question are not clearly outside the reasonably expectable range of chance phenomena. Note that we are *not* insisting that those families are *not* related—one cannot ever prove that, because it is always possible that related languages have been diverging for so long that the inherited similarities between them have become too sparse to support a solid inference of the relationship; but we do say that relationships must be proved and that the Nostratic hypothesis has not been proved (Nichols 1996:52–4; Ringe, forthcoming). I am not optimistic about its prospects.

Finally, since reliable shortcuts are still being worked out, what does traditional, rigorous historical linguistics have to offer the archaeologist or the population geneticist? Amazingly little, in fact. There are quite a few reasons for that—for example, the well-known difficulty of correlating linguistic and archaeological patterns of data in situations in which there are no internal links between the two. I will here discuss two reasons why linguistics isn't very helpful which are often overlooked, not because they are arcane, but because they are so simple.

In the first place, we should consider what events would have to occur to significantly change the genetic character of a whole population. Not only would a fairly large influx of foreigners (relative to the size of the original population) be necessary, but it would take many generations of interbreeding to scramble the gene pool thoroughly. On the other hand, a community can abandon its old language and adopt a new one much more quickly than that—within a single century, in fact—and communities do that with considerable frequency; even stone-age communities have been observed to shift languages (in highland New Guinea, for example, or in northeastern North America during the Iroquois wars). Of course the language-shifters do have to have some native speakers of the new language from which to learn it, but if the latter community is powerful enough (politically or economically), it can be fairly small; good examples are the Latin-speaking communities of Roman Gaul, whose language eventually supplanted Gaulish, or the English ruling class of the Danelaw after its reconquest by Alfred's descendants, who managed to impose their language on a farming population that was largely Scandinavian in some areas. By now the reader can foresee my conclusion: we do not expect genetic boundaries and linguistic boundaries necessarily to coincide. If they do coincide, that is extremely interesting and undoubtedly significant; but if they do not, their failure to do so need not mean much.

Secondly, there are severe limits on the time depth to which we can recognize linguistic relationships. Recall that every language family was once a single language, and each such language must have been spoken by real people at some real time and place; but it turns out that if the time in question was more than about ten millennia ago, we won't even be able to recognize that the descendants of that prehistoric language really are a family. The reason for this is simple: we can only recognize families by finding regular sound correspondences between the languages, and we can only find regular sound correspondences to the extent that the languages in a family preserve the original words and grammatical affixes of the language from which they are all descended. But of course languages often do not preserve their inherited words and affixes; in fact, one universal type of language change is the replacement of old linguistic material by completely different words and affixes (from a variety of sources). As it happens, even the most basic words and affixes get replaced by totally new ones distressingly fast (on the average), so that after 10 000 years (at most) what is left of the parent language in two or more of its daughters cannot be distinguished from chance resemblances. Thus there is a ceiling on how far back into prehistory we can work; and though it varies somewhat according to circumstances, there is absolutely no way around it. Moreover, the ceiling is low, relative to the probable antiquity of human language (which has surely been around for much longer than ten millennia). Consequently, although most linguists of my acquaintance believe that all known human languages have diverged from a single prototype, we have no hope of proving that, let alone of reconstructing any part of that ultimate protolanguage; we must be content to discover language families that began to diversify from individual languages within the past few millennia, and to reconstruct those parent languages of the relatively recent past. It follows that we have nothing to offer anyone investigating the events of human prehistory before the Neolithic revolution of the Middle East—and that is most of human prehistory.

I am overwhelmed by a sense of irony every time I have to insist to nonlinguists that my field is not actually able to perform the miracles they may have expected. But I do believe in telling the truth, and I think the truth is that historical linguistics has been oversold as a window into remote prehistory. However, it is not mainstream linguists who have oversold it; it is the detractors of mainstream historical linguistics who have been trying to convince the public that linguistics can work wonders. More judicious colleagues have insisted on much greater caution. We trust that an intelligent public will be properly sceptical of surprising claims about linguistics and prehistory, no matter how they may offer to satisfy the universal desire to know more about our origins.

<h2 style="text-align:center">NOTES</h2>

1. I would like to thank President Sir David Smith for doing me the great honor of inviting me to speak as part of the Wolfson Lectures, Professor Anna Morpurgo Davies for reading an earlier draft of this paper and making extensive suggestions for its improvement, Dr. Bryan Sykes for much helpful editorial feedback, and Dr. John Penney and the lecture audience for helpful comments. Any remaining errors or other shortcomings are entirely my own.
2. This expresses the real-world situation more accurately than the traditional statement that the languages of a family are 'descended from' a single language.

3. No hard-and-fast line can be drawn objectively between 'languages' and 'dialects'. As a rule of thumb, linguists treat as dialects of a single language two or more forms of speech that are mutually intelligible (sometimes only after fairly brief exposure and a certain amount of practice); mutually unintelligible forms of speech are assigned to separate languages. There are, of course, intermediate cases of various kinds, and they must be described objectively as we find them; nothing is gained by forcing them into a descriptive straightjacket according to which they must either be separate languages or not. This can seem unsatisfying (or even upsetting), because it is only natural to invest the notion 'language' with political, social, and cultural significance; but a culturally charged view of the matter is exactly what linguists wish to avoid, because it renders objective description and analysis impossible.

4. In other words: all the Romance languages are divergent developments of Latin; Latin and various other ancient languages were in turn divergent developments of Proto-Indo-European. Thus Romance can be treated as a family in its own right, but in a larger perspective it is a subfamily of Indo-European. The situation of the other families named is analogous: Arabic is a subfamily of Semitic, which is in turn a subfamily of Afro-Asiatic; Han is a subfamily of Sino-Tibetan.

5. Historical contact between languages that are already quite different from one another can give rise to new similarities between them, of which the simplest and most obvious is the borrowing of words from one language into another by adult speakers; in addition, languages resemble one another because they share various universal properties, and individual words and grammatical items of different languages can resemble one another by sheer chance. All these other types of similarities have nothing to do with the classification of languages into families (except that in some circumstances they can make the family relationships much more difficult to discover).

6. For an interesting illustration of the consequences of taking this seriously, see Nichols 1990.

7. For social dialects, at least, other factors come into play, such as the desire to speak like one's peers *and not* like an outsider. In such cases the persistence of social distinctiveness has the same sorts of effects as lack of linguistic contact.

8. Note that a 'line of descent' need not be a separate 'language', in the sense explained in note 3; any two dialects which have not regularly

been in intimate contact, and which thus have not exchanged large amounts of linguistic material, reflect separate lines of descent from a unitary prototype. This can be illustrated even from the history of English. The first sound change described immediately below, the rounding of /ā/ to /ɔ̄/ in Middle English, occurred only in the more southerly dialects; in the northernmost dialects /ā/ remained unchanged. The result was a regular sound correspondence (see below for an explanation of this term) between northern /ā/ and southern /ɔ̄/, precisely as if those dialects had been separate languages.

9. Determining the (approximate) pronunciation of languages of the past is an exceptionally technical and arcane subfield of linguistics, requiring a thorough knowledge of phonetics (i.e., the physics of speech) and linguistic sound structures and a wide array of specific information about the particular case in hand. How to use the different types of information available is perhaps best explained and exemplified in Sturtevant 1940; good modern examples are Allen 1978, 1987. For the pronunciation of Old and Middle English no similarly convenient handbook exists; the results of such investigations are summarized in Brunner 1965, Campbell 1959, Mossé 1952, and other standard grammars, while Kökeritz 1961 is a handy summary for the dialect of Chaucer. For later stages of English a great wealth of sources is available. Particularly accessible is Jespersen 1948; of the more specialized works, Kökeritz 1953 is especially interesting because of its subject, while Wolfe 1972 includes an exceptionally close scrutiny of a wide range of sources. All these works presuppose at least a rudimentary knowledge of phonetics; discussing speech sounds objectively without a basic understanding of phonetics is not possible. The phonetics taught in a good general introduction to linguistics should be sufficient to make any of the works cited here intelligible.

10. Here and below I use my own pronunciation of standard American English. An excellent description of the differences between standard pronunciations of English in various countries is Wells 1982. Since it is the sounds of language that matter, not the spellings, linguists customarily transcribe languages into a phonetic alphabet. Sounds which are meaningfully different from each other in the sound-system of a single language ('phonemes') are written between slashes; which differences are meaningful is not the same, of course, from language to language, or even from dialect to dialect, so that

each language's or dialect's system of phonemes is idiosyncratic and unique. I will use this objective system of notation only where the exposition might be insufficiently clear without it, so as not to burden the reader with unnecessary apparatus.

11. In this and similar tables '>' should be read as 'developed into'. The OE letter þ ('thorn') represents an interdental fricative—voiced in some phonetic environments, voiceless in others—which usually survives with little change in Modern English.

12. The *w-* of this word is pronounced in some dialects, in which it arose by another sound change, but it seems never to have been pronounced in the London dialect, the ancestor of the standard dialects of Modern English (cf. the entry 'wh' in the *Oxford English Dictionary*); thus for standard English this *w-* is completely bogus. English orthography is unfortunately full of such irrationalities.

13. See note 10.

14. The length of vowels in Latin words is known in remarkable detail, partly because the structure of classical Latin verse depends so heavily on it. The stress pattern is known from the discussions of ancient grammarians. Interested readers should consult Allen 1978.

15. Latin nouns and adjectives are given in the accusative case, the etymological source of the western Romance forms.

16. The *d* of this word has never been pronounced; it was introduced, in writing only, because French *poids* translates Latin *pondus*.

17. The claims made in Greenberg 1987 will be discussed below.

18. See the long discussion at the beginning of Chapter 1 of Greenberg 1987, in which he overestimates the size of the task by maintaining that the standard comparative method (or any other exact method) must involve *pairwise* comparison of *all* pairs of languages in which the linguist is interested. That might have been true if the relationship between every single pair (assuming there were any discoverable relationship at all) were so distant as to be far from obvious. But we typically find a good many fairly large sets of languages, like those from which the examples above were drawn, whose close relationship is clear from the start and whose main sound correspondences can be worked out rather quickly, vastly reducing the size of the overall problem. Moreover, it is clear that genetic relationship is a transitive property (since it is neither more nor less than descent from a single language of the past), so that if we can show that language *x* is related to one language of an already established family, it will follow that it is related to the whole family; thus to

relate Yoruba to the Bantu group we need only show that it is related to one Bantu language (or, still better, to reconstructed Proto-Bantu, the parent of the entire group)—not to all 400 or so modern Bantu languages individually. Every working historical linguist knows this.

19. I have taken my examples from more than one 'typical' Algonkian language because all have lost some of the words inherited from Proto-Algonkian. In these tables the colon indicates that the preceding vowel is long, the accent marks over Arapaho vowels indicate tones, and '<' means 'developed from".

20. Cf. also Ojibwa *tipikkat* 'it is night'.

21. This is not a perfect cognation, because the Arapaho possessive prefix *n- has been remodelled by analogy with other possessed forms; that is why the vowels of the first syllable don't match. As the reader will by now have gathered from various hints, language change isn't all regular sound change; see the end of this paper for an especially disruptive consequence of that fact.

22. Exactly how rare such events are depends on our answer to the question 'One in a million *what?*'; that is, it depends on the universe over which the probability is calculated. Here the proper context seems to be sets of words that translate each other in any five languages; and since there are about 6,000 known languages, giving $\binom{6000}{5} = 6.469 \times 10^{16}$ sets of five, and each language has a vocabulary of several thousand words, it is clear that chance similarities on the scale we are talking about are not, in fact, impressively rare. They will of course be rarer if we restrict the scope of our investigation to some list of 'basic' vocabulary and to some subset of the world's languages (say, those of Australia).

23. In this and subsequent tables the figures in the two columns are rounded independently.

24. It might reasonably be objected that I am looking only at distributions of initial consonants, whereas one ought to consider the rest of the word as well. That would be a serious objection if most of the comparanda adduced in long-range comparison were several phonemes long and the correspondences between the phonemes were very precise; but in my experience that is not how 'long-rangers' operate. For example, in the 'Amerind' dictionary (on which see further below), typically only a single consonant is comparable across the whole set of putative cognates adduced; with very few exceptions, any further similarities are very approximate and/or

restricted to relatively few of the languages. For an illustration of how to compare whole word-forms in such a way as to find greater-than-chance resemblances in unpromising material see Ringe 1996:139–40.

25. I should perhaps emphasize that it is the methodology of long-range comparison that is under discussion here; I here use Greenberg's work as an example because it is prominent, easily accessible, and widely used by nonlinguists.

4

Human evolution: our turbulent genes and why we are not chimps

Gabriel Dover

We might not know why we are the way we are, but the one thing which is certain is that we sometimes regard ourselves, *Homo sapiens sapiens*, as qualitatively different from the rest of the living world. The species-centric question that would arise in most minds is, I believe, 'why is our species different from all other species?', rather than considering the more modest question of why we are not chimps. The more expected approach reminds me of that other Big Question traditionally asked by small Jewish children at the start of the Passover ceremony as the family sit round a table laden with exotic food and general abundance: 'Why is this night different from all other nights?' This innocent query then sets in motion a two-hour spiel from the head of the family on Jewish history, the enslavement in Egypt, and the miraculous escape by divine intervention. All of which is wrapped up in the notion of the 'chosen people'. This sort of heavy-duty answer is not what the child had in mind, who simply wanted a quick answer to why there is so much to eat, and can we please get on with it.

Is there, then, a quick answer to why the human species might be different from all other species? Two sorts of quickie come to mind. The first is divine intervention and the notion of the chosen species; and the second is the origin of the gene (or genes) responsible for, say, our human-specific language facility. The first is not a matter for scientific enquiry, although why so many humans believe in such an uninspiring answer and why some scientists publicly gnash their teeth over the many who do so, is a serious matter of scientific enquiry in itself. The second quick answer centres on the supposedly one human attribute that we can genuinely call our own which, if it could be solved, would allow all the rest of our genetic history to fall neatly into place. One of Sydney Brenner's better jokes encapsulates this position perfectly: if we were to pair off every chimp gene with its human equivalent then one gene would be left unpaired—the language gene. If it's on the human side then we could call it the 'Chomsky' gene; whereas if it's on the chimp side, it could be called the chimpsky gene—and chimps had the sense to keep their mouths shut about it!

I do not believe that the human species can be so simply circumscribed. Although the philosophers of mind, such as John Searle and Jerry Fodor, might disagree with each other over the root causes of human consciousness, they do, nevertheless, agree that there is an inherent 'self-awareness' property of the human mind, spilling over into other attributes of humour, angst, aesthetics and ethics, which are not present in our great ape cousins. It is not obvious that a gorilla, sitting peacefully at the top of a tree and looking out at the sunset, after a good day's foraging and sex, is thinking 'that was a good day, the sky is beautiful and may tomorrow be the same'. I prefer the view that gorillas, and other animals, live in what Gerry Edelman has called the 'eternal present'—that is, they have no deep awareness of the past which can lead, in turn, to a planned awareness of the future. Humans, on the contrary, cannot escape the sentience of their own past, present and future existence. Such potential entrapment is probably unique to humans, as is language, and a genetic contribution to these aspects of our phenotype is hidden somewhere in our genome.

At the risk of instantaneous contradiction of my last sentence, I will state here that it is my belief that there are no specific genes *per se* for language, or for self-awareness, that have evolved *de novo* for such purposes, and which are not to be found in other life-forms. Rather, as with most other evolved novelties, such as the eye, wing, heart, etc., evolution is about teaching old genes new tricks:

it is largely an incessant process of new combinatorial interactions between established, autonomous and redundant modules of biological information. Once a basic set of modular units and developmental operations had evolved, possibly predating the inception of multicellular organization at the time of the Cambrian, it has not been difficult to achieve increasing levels of biological sophistication, without the need to resort to the creation of vast armies of additional genes. 'Complexity', including the assumed complexity of the human brain, is not as intractable as was once thought, requiring the application of chaos theory to unmanageable numbers of genes and their interactions, as championed by Stuart Kauffman.[1] Language and self-consciousness were, probably, the accidental, yet inevitable, outcomes of new genetic interactions which are themselves consequential on the inherent flux in the numbers and genomic positions of relatively small genetic modules (DNA motifs) both within the genes and in the critical regions adjacent to the genes which regulate the place and time of their expression during development.

Novel biological functions can arise from new combinatorial interactions at all levels of biological organization: an insight developed originally by Antonio Garciá-Bellido, but which is now widely accepted. Hence, there are no programmes, or blueprints, within our genome acting as specific sets of instructions leading to the development of a heart or consciousness or a language facility. Species-specific ontogeny is as much a time-dependent, contingent unfolding of the expression of genes and their ensuing interactions on the timescale of an individual, as phylogeny has been an historical unfolding of unique events on an evolutionary timescale. In these terms, the gene has no meaning outside its interactions. In biology $1 + 1 = 7$: the single selfish gene can be neither a unit of function nor a unit of selection. To think otherwise is to turn biology inside out which, whilst superficially clever and seductive, has little meaning in the real world of evolution and development.

4.1 HUMAN–CHIMP DNA DIVERGENCE

So, what can we learn from recent discoveries in the genetics of development and behaviour and in the inherent high levels of flux in the genetic material, about the possible genetic basis of human consciousness and language? Before answering this question it is important to consider some widely quoted data about human–chimp genetic differentiation. The first is that human and chimp DNA, when compared using rather steam-age methods of DNA sequence analysis, involving denaturation and renaturation, turns out to be about 95–99% similar. Secondly, this level of difference, when measured by a supposedly universal clock of genetically neutral divergence, means that we and chimps separated around 5 million years ago from our most recent common ancestor. Despite some degree of circularity in such arguments, a period of 5–10 million years is beginning to hold up from a variety of other studies of the molecular evolution of specific, shared genes amongst the apes. The critical issue is not the precise period of elapsed time but whether there is any justification for the surprise shown by many who cite such data of the supposedly small 1–5% genetic difference. If we give a little thought to the fact that 95–99% of the human genome is probably 'junk' (that is, not composed of long stretches of sequences that can be successfully translated by the genetic code into the corresponding strings of amino-acids), then the surprise would be considerably reduced. The 1–5% difference between human and chimp could encompass all of the roughly 70 000 genes that each genome contains, all of which could be crucially different between the two species. I do not believe this to be the case, but the possibility should temper the speculation.

To return to the question of time: 5–10 million years is not a long time when measured in generations of humans and chimps. In fact the shortage of time to evolve new language-specific, or consciousness-specific genes, becomes even more acute if we start our counting from other significant events in the lineage leading to *Homo sapien sapiens*. We could start counting from the time of the sudden enlargement of the brain at around 2 million years ago.

This would make sense in terms of advanced brain functions such as language and sentience. However, no obvious advancement in technology or culture seemed to have taken place for most of this two million year period. From sequence comparisons of the D-loop of maternally transmitted mitochondrial DNA of modern humans, it has been calculated that 'Mother Eve', and presumably 'Father Adam', lived in populations of true *Homo sapiens sapiens* in Africa around 200 000 years ago. Even were we to stretch this back to 500 000 years, we are still left with very little time to evolve anything of significance at all, on the assumption (which I believe false) that complex new brain functions require *de novo* suites of specific genes.

4.2   DNA FLUX AND REDUNDANT DNA

Before entering into the main story of what we should be expecting to find in our search for the genetic basis of why we are not chimps, a word about the 95–99% junk DNA. We know that this is spread around 23 pairs of chromosomes that are, to all intents and purposes, structurally the same between chimps and humans, even down to the chromosome banding patterns and orders of genes (synteny) along each chromosome. Junk DNA, as in all examined species of animals and plants, is highly redundant with copies that range from two to millions in number and which vary in length from one base-pair to a few thousand base-pairs, [for reviews see reference 2; and papers, reference 3]. The copies can be in head-to-tail arrays (minisatellite, microsatellite and satellite DNAs) or dispersed over thousands of separate loci. Significantly, many families of sequences show a much higher degree of sequence identity between family members within a species than there exists amongst the equivalent members between species. This pattern of mutant distribution is called 'concerted evolution' and has been observed for many repetitive families between humans and chimps (e.g. the alpha-satellite DNA; the 'Alu' family and the ribosomal RNA genes, rDNA). The continual flux in the copy-number of repeats, changes in their genomic positions and the homogenization of families for species-specific mutations (concerted evolution), are due to a variety of non-Mendelian mechanisms of genomic turnover (unequal crossingover, DNA slippage, gene conversion; transposition; retroposition, etc.). Collectively, they underpin the evolutionary process of 'molecular drive' by which new mutant members spread through a family (homogenization) and, ultimately, through a sexual population (fixation) with the passing of the generations. When molecular drive occurs in critically important regions of genes, or amongst the redundant regulatory sites in the promoters of genes, then the phenomenon of 'molecular coevolution' may occur.[4] This is thought to be due to the action of selection promoting compensatory changes in other genes whose products interact with a particular newly homogenized redundant set of sequences, if essential cellular functions are to be maintained during molecular drive. I shall return to this issue later.

What do we know, therefore, of the numbers, types, levels of redundancy and functions of families of repeats that make up the junk DNA of chimps and humans? Very little. This paucity of information is surprising given what we could learn about human evolution were a chimp genome sequencing project undertaken. Maybe we need to wait for some enterprising chimp to make a start. The little we do know is based on years of detailed action-packed studies of the alpha-satellite DNA and 'Alu' family sequences by several world-wide groups of investigators, [see chapters in reference 3]. This is not the place to entertain details of these studies except to say that there are, indeed, in both families, some human-specific subfamilies which have arisen since chimp–human divergence. Such subfamilies arose by unequal crossing-over in the alpha-satellite family and by retroposition in the 'Alu' family. These subfamilies make up a total of a few thousand members only, which is a small percentage of the half million 'Alu' members and the hundreds of thousands of alpha members in each haploid genome. They should not be totally ignored, however, for they can get themselves into positions, unique to each species, which occasionally affect the regulation of specific genes or induce functionally effective genetic rearrangements.

From recent studies on the fine-grained repetition of many species genomes it is apparent that 'DNA slippage' is by far the most active and the most ubiquitous of the genomic turnover mechanisms in generating lengthy tracts of cryptically simple DNA (that is, repetitive short DNA motifs of 1–6 base-pairs which are interspersed one with another). I would expect that the composition of such DNA in homologous regions between chimp and human would be very different. Given that slippage is known to be operative both in the regulatory promoter regions of genes, as well as in the bodies of genes (exons and introns) themselves, some of our essential phenotypic differences of language and sentience might have been influenced by such fine-grained and subtle differences in the combinatorial patterns of repetitive short motifs, accumulating over time.

4.3   MODULARITY AT THE GENETIC LEVEL

The time has come to return to the issue of modularity, redundancy, autonomy, co-option and molecular coevolution—the total phenomenology of evolved biological structures which emerges from the interaction between the external turbulence of the ecology that drives natural selection and the internal turbulence of the genes and their controlling elements that promotes molecular drive.

Modularity exists at all levels of biological organization—DNA, proteins, cellular and developmental processes, and organs. At the genetic level, both gene regulatory regions and the genes themselves are modular. Figure 4.1 illustrates the modular and redundant nature of the promoters of several unrelated genes in *Drosophila melanogaster*, although the genes are all expressed in early embryogenesis. Each module is a short motif of sequence that is recognized and bound to by corresponding motifs of amino acids in the proteins involved with the activation or repression of transcription. Several noticeable features emerge from the genes displayed in Fig. 4.1. First, the copy-number and composition of each module type has altered as one moves from one gene to

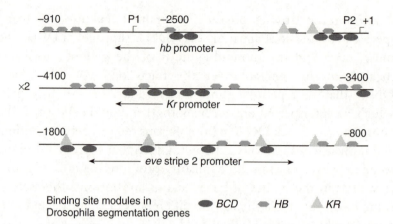

Fig. 4.1 The promoters of three unrelated genes *hunchback (hb)*, *Kruppel (Kr)* and *even-skipped (eve)* are depicted upstream from their respective genes. Each promoter contains a number of target sites to which regulatory proteins bind, repressing or activating the transcription of the downstream gene. The binding sites are symbolized in three ways, as shown, for the three regulatory proteins, BICOID (BCD), HUNCHBACK (HB) and KRUPPEL (KR). Note (i) binding sites are in varying numbers of copies due to genomic flux; (ii) they are shared by unrelated promoters, hence must be mobile; (iii) a gene's level of transcription is a reflection of combinatorial interactions amongst the different numbers and types of modular binding sites (see text).

another. Such changes are known to affect the strength of binding of the regulatory proteins. Secondly, and importantly, module types are shared by unrelated genes—indicating that they must have been mobile by one means or another; probably as a consequence of one or other of the turnover mechanisms mentioned in the previous section.

The shared, modular nature of promoters has been found in a variety of gene regulatory regions where detailed comparisons have been made. Two recent sets of results derived from the experimental manipulation of the modular nature of gene promoters in *D. melanogaster* by Michael Levine and co-workers[5] and in sea-urchins by Eric Davidson and co-workers [reviewed in reference 6] reveal the generally autonomous nature and short-range effects of regulatory modules. This is not the place to enter into the details of such experiments except to point out two essential points. One is that promoters can be mosaics of several different module types whose position, sequence, effects (positive or negative), binding efficiency and synergy with each other affect the time, place and level of gene expression. The second is that the modular nature and short-range effects of promoters introduce an enormous level of combinatorial flexibility, on an evolutionary timescale. New modules, or groups of modules, can transpose into existing sets of modules without necessarily bringing about a major disruption of pre-existing gene expression. Instead, additional places and times of expression during ontogeny can occur, bringing a particular gene's protein into potentially new interactions with other proteins which are expressed at the same times and places. Furthermore, the redundant nature of module types can add a degree of genetic buffering against the effects of mutations in module sequence. Such buffering, when coupled to the ability of repeated sequences to effectively 'cross-talk' (via, for example, gene conversion) can aid in the establishment of 'molecular co-evolution' between the multiple target modules in the promoters and the DNA-binding module of a given regulatory protein. By such means, to be described later, biological incompatibilities can arise between the cis-acting modules (binding sites) of one species and the trans-acting binding modules of another, in addition to the inception of novel molecular permutations within a species.[4]

4.4   MODULARITY AT THE PROTEIN LEVEL

The modular nature of proteins enjoys a long history of study [for reviews see references 7, 8]. As with the promoters, proteins are often mosaics of different module types that can differ in position and number, and which are frequently shared by unrelated proteins (e.g. the homeodomain module—see later). It has been calculated that there may be as little as 1500 modules shared by all of life's proteins. This is a surprisingly small number. The next surprise is that despite the theoretically infinite number of combinations of 1500, given a crude average number of 10 different module types in each protein, the actual number of genes (proteins) is also very small. We are talking numbers of approximately 5000 (bacteria), 8000 (yeast), 70 000 (mammals, including humans). If we take on board the possible octaploid nature of the human genome, then there might be no more than 2000 different genes, responsible for all of our human wonders of language and sentience.

At one level, all that this means is that there has been only one tree of life and that the numbers of different types of modules in promoters, or in genes, or in proteins are all that is currently required for life's diversity. In other words, it has not been essential to occupy the whole of the n-dimensional 'space' of modules in promoters/genes/proteins. Most of the genotypic space lies empty. At another level, the paucity of promoter/gene/protein modular types tells us that new functions arise as a consequence of novel combinations of modules and that many modules must participate in several different higher level functions. I have already stressed the extent of module sharing in the promoters of unrelated genes and in the exons of unrelated proteins. The shared participation of whole genes and their protein products is also a fact of life which needs to be taken on board when considering the origins of biological novelties. It is becoming more apparent that novel functions (language? sentience?) are more a question of teaching old genes new tricks rather than the laborious generation of *de novo* genes through gene mutation on pre-existing sequences. Pleiotropy (the participation of a given gene in multiple, often unrelated processes) is the name of the game. Such inherent genetic flexibility and virtuosity would not be possible without the redundant, combinatorial and mobile nature of autonomous modular constructs.

4.5   TEACHING OLD GENES NEW TRICKS

For example, in a recent review of gene number and function[9] the extensive degree to which genes participate in multiple functions is emphatically underlined. It has been estimated that over 70% of the known genes which are lethal to *D. melanogaster* when mutated are active both during embryogenesis and in the brain. Clearly the degree of functional overlap between such genes must be considerable. Similarly, of 20 000 'enhancer-trap' lines in the same species, established to 'trap' genes which function either during eye development or during brain development, only a small handful were found to be specific for the eye or the brain.

However, the story *par excellence* of genes co-opted for entirely new functions concerns a number of genes, known to be normally responsible for the enzymes involved in very basic cellular metabolism, but which have acquired entirely new functions for the production of different types of crystallin protein in the eye lenses of different animals. Again, this is not the place to provide the details of this significant phenomenon, except to make two points. The first is that different crystallin types are the products of entirely different genes coding for different metabolic enzymes, indicating the underlying virtuosity of proteins. Secondly, the level of expression of a shared gene, either in normal metabolism or in the eye, is controlled by alterations in the gene promoter such that some binding-site modules are used in one function and others are used for the second function. Interestingly, the products of the mammalian gene PAX-6 (the so-called master gene controlling eye development in animals—see later) binds to promoter modular motifs that are known to be different between two different crystallin genes. It is possible that a molecular coevolution has occurred between the specific binding sites of a given crystallin promoter and the binding motif within the PAX-6 protein, as is the case with other transacting proteins and their cis-acting targets (see below). In other cases of crystallin gene co-option, it has been shown that the metabolic enzyme gene has first undergone a gene duplication, presumably by unequal crossingover, followed by divergence in the promoter of each of the duplicates so that one performs the usual function and the other performs the new lens function. For a review of this story and many related issues see reference 10. It is becoming clear that there are diverse ways of co-opting existing genes into new functions involving genomic turnover and the establishment of new modular interactions.

4.6   MODULARITY AT THE CELLULAR AND ORGAN LEVEL

Modularity at the level of gene and protein have been relatively easy to accept, almost to the point that the norm is one of expectation, not surprise. Modularity at the level of developmental processes and organs is a more recent and surprising phenomenon, although the first hints of modularity at the level of organs were perceptively recorded by William Bateson[11] at the turn of the century, leading to his coinage of the term homoeotic (now abbreviated, carelessly, to homeotic) mutations. These are mutations in the genetic material responsible for major transformations of one part of a body to another: for example, the appearance of a complete segmented leg in place of an antenna on the head of insects, or the development of four-wing dipteran flies through the replacement of the small pair of halteres (used for balancing) by a second pair of wings. Not unexpectedly, such homeotic transformations were marginalized by rigid Darwinians as a freak sideshow peripheral to the main act which, by mathematical necessity, involved the gradual selection of minor mutations in armies of polygenes responsible for the form and behaviour of organisms. This dismissal of what we now regard as the raw material of evolution held back any useful cross-talk between students of evolution and students of development for over 80 years. It also led to the failure to recognize the true signifi-cance of homeotic mutations, which is that organisms are con-structed, phylogenetically and ontogenetically, as the equivalents to biological Lego sets. As we have seen with genes and proteins, organisms can be built, starting from a few basic modules, into complex creatures, in which 'complexity' is a reflection of the particular permutations of interactive modules that make up a given species constitution. Hence, legs and antennae share many of the same modular building blocks, for which the appropriate combin-ation during development depends on the activities of a relatively small number of 'selector' genes, as defined by Antonio Garciá-Bellido.[12] A modular leg can be bolted onto a head in place of an antenna during development without recourse to a complete genetic reorganization or restructuring of the head.

Many homeotic genes were found to contain the so-called 'homeobox' of sequences that codes for a stretch of about 60 amino-acids (the homeodomain) within a longer protein, which binds to particular short modules of DNA in the promoters of a number of diverse genes. When the 'homeobox' was first discovered, it was hailed as the Rosetta Stone of biology. Today, it has to take its place more modestly amongst a variety of other long-conserved DNA-binding motifs that exists in families of other gene regulatory proteins (transcription factors). Nevertheless, as explained earlier, the total number of DNA-binding motifs is surprisingly few, given the total diversity of life forms that we are trying to explain.

Recent findings that the so-called 'Hox' gene cluster, controlling segment identities along an anterior–posterior axis in insects, exists and performs similar jobs in vertebrates (including humans) testifies to the very ancient origins of such 'selector' or 'master-control' genes [see reference 13 for reviews; and chapters in reference 14]. Similarly, the ability of PAX-6 (the master-gene controlling eye development in mammals), to switch on an alternative set of genes which make insect eyes, but in diverse and unusual parts of an insect's body, also underlines the modular nature of development.[15]

There is an inherent flexibility in the construction of organisms which arises from the ways in which genetic operations are performed during development: that is, interactive systems of ancient modular components.

4.7 EVOLUTION: MODULARITY AND COMBINATORIAL FLEXIBILITY

Ontogenetic flexibility and the existence of ancient genetic operations, stretching back presumably to the Cambrian, strongly suggest that evolutionary flexibility is based on the same principles. The 'module' of an intact developmental or cellular process has been termed a syntagma (plural syntagmata but sometimes anglicized to syntagmas) by Antonio Garciá-Bellido [for recent account

see reference 16] or a Seminal Regulatory Interaction (SRI).[17]
What is becoming clear is that evolutionary novelty emerges from
the continual shuffling of modules (at the levels of genes, proteins,
developmental processes, organs) and the establishment of new
interactions in new places and at new times during ontogeny.

Hence, the novelty of language or sentience, features which we
can safely assume distinguish our human species, probably arose
like much else in evolution from the fortuitous emergence of
novel interactions between pre-existing genetic and phenotypic
modules. What these might involve is anyone's guess. Given the
extensive pleiotropy (multiple functions) of most genes, it should
not come as a surprise, as with the eye lens crystallin story, that
human superconsciousness and language involve genes with quite
bizarre and unrelated functions, elsewhere in our bodies. That
there are genes involved is not in doubt; however, that such genes
are unique and specifically evolved for the emergence of language
and sentience in this human species, is not a serious proposition.

4.8   DNA FLUX AND MOLECULAR CO-EVOLUTION

If the evolution of novelty in form and behaviour is primarily
about the establishment of new genetic interactions, how and why
are they tolerated during the course of individual development?
The secret here lies in the dual interlocked phenomena of redundancy and molecular co-evolution between functionally interacting molecules.

As I explained above, there are a variety of ubiquitous mechanisms of genomic turnover that place the genetic material in a
considerable state of flux, on an evolutionary timescale. Such
mechanisms generate a high degree of genetic redundancy involving multigene families, internally repetitious regions within genes,
multiple copies of regulatory binding sites in promoters and, of
course, vast amounts of repetitive junk DNA. It is often the case
that, after a mutation has arisen in one member of a redundant
genetic element, then that mutant member can replace all other
members, either as a consequence of the continual, stochastic,

processes of gain-and-loss by, for example, unequal crossingover and unbiased gene conversion, or as a consequence of more directional processes such as biased gene conversion. Elsewhere I have described the dynamics of spread of mutant elements in redundant families.[2] In brief, as a consequence of the large disparity in rates between (i) the rate of mutation generating a mutant member in the first place, (ii) the rate of genomic turnover homogenizing the mutant member through a family, and (iii) the rate at which sex shuffles chromosomes amongst individuals at each generation, then the change in the genetic composition of a population, with respect to the homogenization of a given redundant family, is gradual and cohesive. That is, all individuals at any given generation of a sexually reproducing population are, more or less, in the same boat. An important consequence of this is that if, say, a redundant set of cis-regulatory binding sites is gradually changing in both copy-number and sequence, in unison, amongst all individuals of a population, then there is time for selection to promote compensatory mutations in the genes whose transacting products need to interact with the cis-binding sites. Hence, the interaction between natural selection and molecular drive ensures a 'molecular co-evolution' between genetic elements that need to interact to maintain essential developmental and cellular functions.[4] Genetic redundancy provides the relaxed conditions for such co-evolutionary changes to occur and be absorbed into the overall developmental and cellular processes. It provides a degree of internal tolerance by allowing interacting molecules to adjust one to another, through compensatory mutations; and on occasion to tolerate novel molecular combinations leading to novel phenotypes.

4.9 BIOLOGICAL TOLERANCE AND THE INTERNAL 'TANGLED BANK'

It is my belief that the primary concern of an organism in solving problems generated by mutation and turnover is to ensure that a co-evolutionary partnership continues between consenting molecules in subtly varying permutations. The main criterion, on which the decision-making process is based, is to ensure that a successful ontogeny can proceed; that is, to ensure that a reasonably functioning organism is being produced without constant reference to the usual external parameters that govern natural selection. Slightly modified ways of doing ontogeny are not necessarily and constantly being checked by the ecology but, rather, the acceptance of such modifications depends on the maintenance of genetic crosstalk between interacting molecules throughout a period of change. Genetic redundancy ensures that there is, inevitably, a more relaxed and tolerant relationship between interacting molecules and a more relaxed and looser relationship between

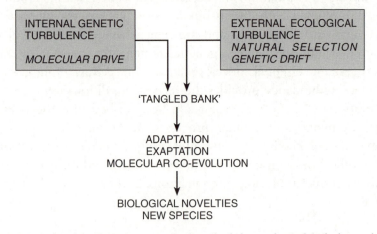

'It is interesting to contemplate a tangled bank, clothed with many plants of many kinds, with birds singing on the bushes, with various insects flitting about, and with worms crawling through damp earth, and to reflect that these elaborately constructed forms, so different from each other, and dependent upon each other in so complex a manner, have all been produced by laws acting around us.' *The Origin of Species*

INTERNAL GENETIC TURBULENCE

MOLECULAR DRIVE

EXTERNAL ECOLOGICAL TURBULENCE
NATURAL SELECTION
GENETIC DRIFT

'TANGLED BANK'

ADAPTATION
EXAPTATION
MOLECULAR CO-EVOLUTION

BIOLOGICAL NOVELTIES
NEW SPECIES

Fig. 4.2 In his day, Darwin correctly described the ecological turbulence that governs natural selection as a 'tangled bank'. Today, we also recognize an internal genomic turbulence that governs the evolutionary process of molecular drive. The true 'tangled bank' is, then, a mixture of the internal and external causes of turbulence. The evolutionary solutions to the problems arising in such a 'tangled bank' lead to adaptations, exaptations and molecular co-evolution which contribute to the origins of biological novelties and new species (see text).

organisms and their environment. At both levels, internal and external, the new biology of modularity, redundancy, genetic turnover and molecular co-evolution force us to abandon the hard lock-and-key imagery of evolution driven solely by natural selection: for example, the standard image of enzymes locked into their substrates, and organisms locked into their niches through the billion-year-old processes of honing and refinement supposedly forced onto biology by natural selection acting alone. Organisms are a consequence of three major forces in evolution (natural selection; neutral drift; and molecular drive): it is no longer feasible to describe the evolution of biological novelties as a response solely to the forces envisaged by Darwin in his famous image of the 'tangled bank' (see Fig. 4.2). There is as much a 'tangled bank' within our genomes as there is in the ecology. Biology's ability to handle the internal 'tangled bank' is based on the tolerance and flexibility that ensues from modularity and redundancy.

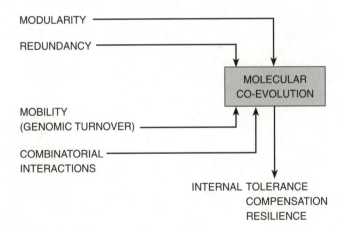

Fig. 4.3 Molecular co-evolution between functionally interacting molecules is a consequence of modularity, genetic redundancy, genomic turnover and combinatorial interactions at the heart of cellular and developmental processes. Molecular co-evolution (essentially the co-evolution of compensatory mutations) reveals that biological organisms are more internally tolerant and resilient to evolutionary change than is expected from a strictly selectionist 'lock-and-key' view of evolutionary success (see text).

4.10 KEEPING THE PLANE FLYING

Biological evolution has often been likened to the human-made changes in the development of the aeroplane, from the first wire-and-string contraptions to the modern supersonic jets. There is, however, an essential difference in that biology does not have the luxury of bringing the planes down to earth every time it wants to initiate a change. The secret of biological evolution is to tolerate major changes whilst the thing is still flying in the air. We need to look at the internal molecular workings of organisms, at the redundancy and flux in the genetic material, and at the modular process of ontogeny to understand how this is achieved. When that has been solved, the discovery of the genetic origins of language and sentience will come as no surprise.

5

Using genes to map population structure and origins

Bryan Sykes

5.1 INTRODUCTION

I want to begin my paper at a familiar landmark. This famous diagram shows an evolutionary relationship between genes and language (Fig. 5.1). On the right, proto-languages divide into language families which in turn disperse into the many different tongues we know today. On the left, we have what looks like an evolutionary tree of human populations. At first glance they look like mirror images of one another—two perfectly matching structures that dispel any lingering doubts about the co-evolution of genes and language. When I first saw this diagram, any misgivings I might have had about the left hand side, the genetics, about which I might be expected to know something, evaporated when I saw the reassuring support of another discipline, about which I know next to nothing. Yet we have heard from Professor Ringe (see Chapter 3) that there are serious reservations among linguists about the methods used to construct some of the language relationships. If that side is fragile, what about the other? Rather than admiring a wonderfully robust construction built on the secure foundation of two independent disciplines, are we instead looking at two tottering piles which, in an almost physical sense, must lean on each other to prevent a mutual collapse. Professor Ringe has already given his view of the construction of the linguistics component. My purpose is to take a closer look at the architecture of the buttress built around the genetics.

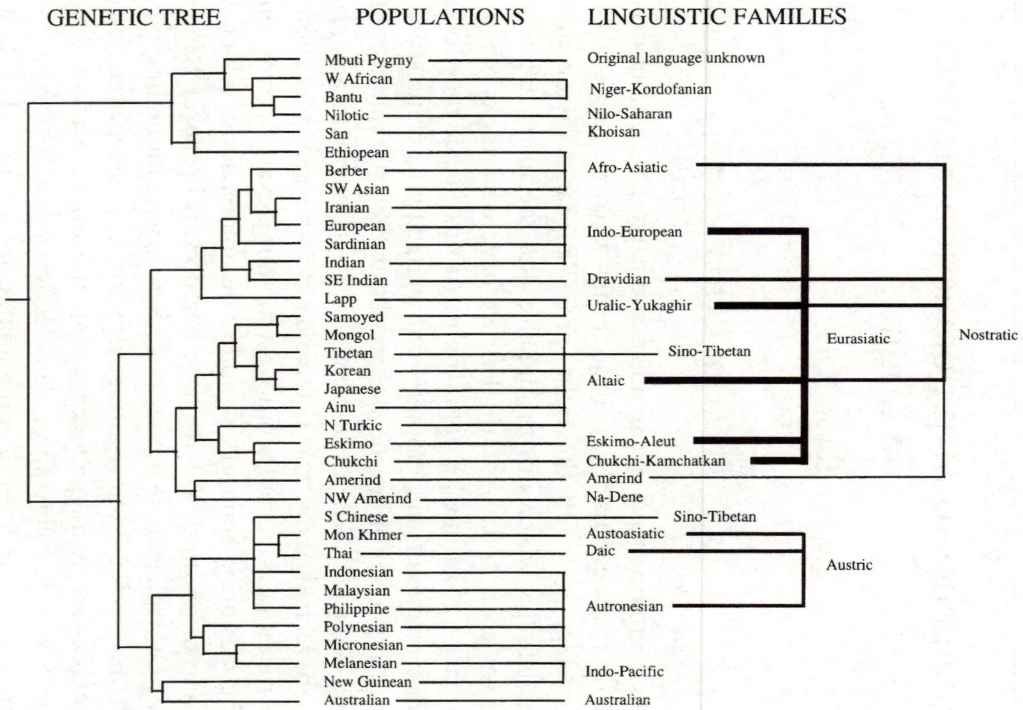

Fig. 5.1 The relationship between languages and genes. On the left is the 'genetic tree' of the different populations listed in the centre. To the right, different language families with some of the evolutionary relationships between them.

To find the first attempts at using genetics to study the relationships between populations, we need to go back to the First World War, to a paper published in the *Lancet* entitled 'Serological differences between the blood of different races—the results of research on the Macedonian Front'.[1] To give you a flavour of the sort of thing the *Lancet* published in those days, the article was sandwiched between a discourse by the eminent surgeon Sir John Bland-Sutton on the third eyelid of reptiles and a War Office announcement that those nurses who had been 'Mentioned in Despatches' for their work in Egypt and France would soon be getting a certificate from the King showing his appreciation!

Because, as now, blood groups were widely tested for transfusion cross-matching there was a mass of data being accumulated during the Great War. Ludwik Herzfeld had, some years before, already established that the ABO blood group system discovered by Landsteiner in 1901 followed the basic genetic rules first formulated by Mendel. His wife Hanka, as head of the Central Bacteriological Laboratory of the Royal Serbian Army, had at her disposal the data from Allied soldiers from many different parts of the world. Working together, they noticed considerable differences in the frequencies of blood groups A and B in soldiers who came from different 'races'. For obvious military reasons, they didn't have the German data to hand, and the figures published in the *Lancet* were 'quoted from memory'!

Even though this was the very first paper of its kind, I'm glad to report there was no shortage of imaginative interpretation. According to the Herzfelds, there were two 'biochemical races' of humanity: Race A, with blood group A, and Race B, with blood group B. Because Indians had the highest frequency of blood group B they conclude 'We should look to India as the cradle of one part of humanity.' As to the mechanism of spread they go on 'Both to Indo-China in the East and to the West a broad stream of Indians passed out, ever-lessening in its flow, which finally penetrated to Western Europe.' They weren't sure about the origin of Race A and thought it might have come from somewhere around North or Central Europe. We know now that this is nonsense but it does illustrate that geneticists, then as now, are never shy of speculation. Because of the wide availability of frequency data through blood transfusions, by the 1950s substantial world distribution maps had been compiled. The Herzfelds were indeed correct that the highest frequency of blood group B is found in India and Tibet but they would have had to revise their view on the origin of the second part of humanity away from Northern and Central Europe to somewhere north of Denver, Colorado because that's where the frequency of blood group A peaks.

I have taken the liberty of drawing a tree from the First World War data in the same way as the gene-language diagram was constructed (Fig. 5 2). The French and Italians have very close values for the frequency of blood group B. The English are slightly further away and the depth of their separation on the diagram is a measure of the 'genetic distance' between these populations. However, some quite odd things are happening. It comes as no great surprise that the French and Italians are genetically similar and both the Germans and Greeks also fit quite nicely into a European group. But what about the Russians and Malagasy. They are genetically identical according to this tree. Are we picking up the echoes of a long forgotten Russian settlement of Madagascar—or vice versa? There are other surprises. Negroes are almost as close to Russians as we are to Greeks, which seems a bit unusual to say

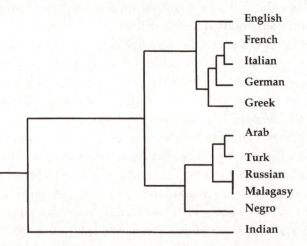

Fig. 5.2 A 'genetic tree' constructed by the same method as Fig. 5.1 from the blood group frequency data of Herzfeld and Herzfeld.

the least. By taking just one genetic system, the ABO blood groups, the analysis produces what appear to be some very reasonable comparisons between populations. Others, however, look distinctly odd.

For the next forty years, huge amounts of data accumulated for ABO and the other blood group systems, like Rhesus, but still the number of strange bedfellows that were thrown up was disturbing. William Boyd, who led the field during the 1930s and 1940s actively discouraged anthropologists from taking any notice of blood groups. One wrote to him—'I tried to see what blood groups would tell me about ancient man and found the results very disappointing'.[2] William Boyd's declared motives for looking at the genetic difference between races would be music to modern liberal ears. He writes, in the same paper: 'In certain parts of the World an individual would be considered inferior if, for instance, he has a dark skin but in no part of the World does possession of a blood group A gene exclude him from the best society.'

As we have seen, there are some very peculiar population relationships that spring up by this kind of comparison between frequencies at a single genetic system. The next advance came in the 1960s when Luca Cavalli-Sforza and his colleagues worked out a way of performing a statistical amalgamation of the published data from many different genes. This averaging process certainly does remove a lot of the more eccentric connections. Nonetheless, the results are still displayed in the same format, as a diagram of genetic distances between populations. At first glance, this looks like a tree of evolutionary relationships. A proto-population appears to split into two populations which themselves split into further subdivisions and so on until the diagram is complete.

The rationale for reconstructing these evolutionary relation-ships is that the genetic characteristics of two sub-populations, as defined by the gene frequencies, diverge away from each other as they both evolve independently. The longer this process goes on, the more different the frequencies become and so the 'genetic distance' increases. However, Cavalli-Sforza and his colleagues clearly do interpret these trees as a history of population splits, as an evolutionary history of human populations. The problem is that, despite appearances, these are not evolutionary trees at all. They are not, in the language of taxonomists, *cladograms*, where the entities at the ends of connected branches, the taxa, have by definition derived from a common ancestor. They are instead *phenograms*—diagrams of overall similarity with no condition of a common ancestry between connected taxa. To quote the author of an elementary taxonomy book 'It is probably unreasonable to expect the majority of people to interpret phenograms as anything other than suggestions of evolutionary history'.[3] If human evolution really were a history of population fissions with no mixing after the splits, with sub-populations of the same size to equalise the effect of drift[1] and with no selection to reduce or maintain diversity, then they could give similar patterns. There might conceivably be scenarios, the colonisation of the Americas being one example, where there may have been little mixing after the original split, where the phenogram might have some evolutionary relevance. However, in others, like Europe, the necessary premises are clearly unreasonable. Can one really imagine a proto-population splitting into the English and the French without any subsequent mating contact between them?

Another necessary evil of the method is that the unit of comparison, the thing at the ends of each branch, is the 'population'—which it requires us to define. Even as long ago as the 1950s, the intellectual hazards of doing this were appreciated. In 1954, H. J. Fleure, referring to work on individuals writes 'it must increasingly displace the study of population groups (being) treated as units in spite of their mixed nature.'[4] As one might expect, the language sharpens, the style deteriorates and the political overtones begin to emerge in more recent comments such as that from Zegura and his colleagues writing in 1990: 'Objectively defined races simply do not exist—the quintessentially arbitrary and subjective race concept is moribund as a unit of human biological analysis'.[5] It strikes me that the numerical approach is intellectually and politically suspect for the simple reason that frequencies have to measured in groups (individuals don't have frequencies) and any conclusions will be generalisations about this or that group. The difficulty that the Human Genome Diversity Project

has in attracting widespread political support may in some measure
be due to the failure to appreciate these limitations.

5.2   MITOCHONDRIAL DNA

Fortunately, there is an alternative approach and that is what I
want to concentrate on for most of the rest of this paper. This is an
entirely new way of looking at genetic histories. The new era, and
that is not an exaggeration, was heralded just over 10 years ago, in
January 1987, when Rebecca Cann, Mark Stoneking and Alan
Wilson published their famous paper in Nature entitled
'Mitochondrial DNA and human evolution'.[6] The well-known
illustration (Fig. 5.3), which has been elevated to 'one of the classic
icons of Palaeolithic archaeology'[7] is the world-wide mitochondrial DNA tree. Unlike the ambiguous distance diagrams this
really is an evolutionary tree—a true cladogram. It is an attempt at
a reconstruction of an evolutionary history, not of populations nor
of individuals but of a gene. The gene is mitochondrial DNA
(mtDNA for short), whose unusual and useful properties I will
explain in a moment. At the end of each branch is a particular
version of the gene defined by part of its DNA sequence. The tree
shows a reconstruction of the evolutionary relationships constructed on the so-called parsimony principle whereby versions differing by a few DNA sequence changes are shown sharing a
common ancestor more recently than versions differing by many
changes. It sounds like a similar principle to that used to build the
distance trees which I have only just finished criticising. In many
ways it is; differences in mtDNA will accumulate over time from a
common ancestor just as sub-populations diverge away their
common ancestor—the 'proto-population'. But there is one
crucial difference. First of all, we can be absolutely sure there *was* a
common ancestor for any two sequences—and, as we have already
discussed, this is not a universally tenable premise for populations.
So it is intrinsically reasonable to draw an evolutionary tree for
mtDNA whereas it is not for populations.

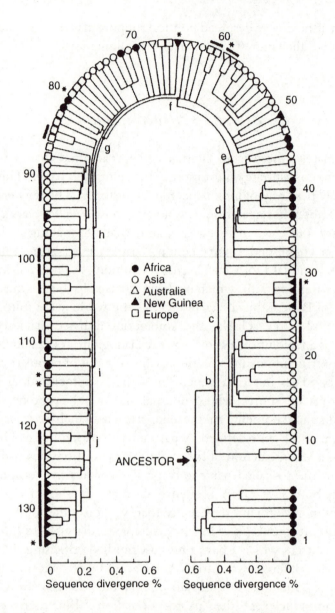

Fig. 5.3 The first mitochondrial gene tree. Around the outside are individuals with their ethnic affinities symbolised. The evolutionary reconstruction of their maternal relationships, deduced from their mtDNA, is shown by the inter-connecting lines. The deepest split from the common maternal ancestor divides one branch that is entirely African from the rest of the world. Note, though, that many Africans (•) cluster with the rest of the world. (From Cann *et al.*[6])

Turning to the detail of the diagram, the symbols at the 'tips' of the branches denote the geographic origin of the individuals who provided the samples. One immediate feature is that there is a deep division between mtDNA from Africans and non-Africans. But even this is not a completely clear-cut distinction. There are Africans popping up all over the tree. Notwithstanding these apparent contradictions, there is pretty obviously an important difference between Africa and the rest of the world. Another feature is that there is a single common ancestor to the whole tree. This almost melodramatic conclusion is actually a logical inevitability but, in the true spirit of theatre, the individual who possessed the ancestral mtDNA sequence was soon dubbed 'mitochondrial Eve'. And, from the diagram, Eve was an African. So Wilson's conclusion was that humans evolved in Africa and spread, within the last 150 000 years or so, to the rest of the world. The time estimates were arrived at by inserting a mutation rate into the calculations—a rate which subsequent work established as reasonable.

The impact of this publication was dramatic. It came down very firmly on one side of the argument about deep questions of human evolution, offering support to the idea, first proposed by Chris Stringer on the grounds of fossil morphology, of a comparatively recent African origin for modern humans. Recent, that is, compared to the alternative view, the multi-regional model, championed by Milford Wolpoff that sees modern populations: Europeans, Asians, Australians, Africans and so on, as directly descended from *Homo erectus*, which emerged from Africa something like two million years ago. Multi-regionalists believe that Neanderthals were the ancestors of modern Europeans and that modern Chinese and Australians are descended directly from Peking Man and Java Man respectively. The mitochondrial gene tree introduced an objective time-depth measurement, based on the mutation rate, and shows quite clearly that the time to the most recent common mitochondrial ancestor of all modern humans is very much less than the two million years required from the multi-regional model. Though its proponents, being thoroughly modern humans themselves, understandably refuse to give in, the mitochondrial gene tree dealt a heavy blow to the idea of multi-regionality from which it has not yet recovered.

I want now to explain rather more about mitochondrial DNA and then to talk about some of our own work, not on this grand, global scale, but rather more locally. Mitochondria are small cellular particles, or organelles, which are located in the cytoplasm that lies between the nucleus and the outer membrane of cells. Their main function is aerobic energy production and they have all the necessary enzymes to do this. They also contain a small amount of their own DNA which carries the genes for some, but not all, of these enzymes. Mitochondria almost certainly began life as free-living, single-cell organisms but, many millions of years ago, entered into a symbiotic relationship inside the cell of another primitive organism and have stayed there in its many descendants ever since. The mtDNA, which is circular and 16 569 bases long (in humans) is tiny compared with the 3000 million bases of the nuclear genome but it does have several useful features for studying evolution. Firstly, mtDNA is inherited maternally for the very simple reason that mitochondria are found outside the cell nucleus. Eggs are comparatively large cells, full of cytoplasm and mito-chondria. Sperm, on the other hand, have only very few mitochondria which, even if they were to get into the egg during fertilisation, are soon destroyed. This means that, though a fer-tilised egg has nuclear genes from both parents in equal amounts, all the mitochondria come from the mother. The embryo, foetus, baby, child and eventually adult, having all developed from the fertilised egg, have maternally derived mtDNA in all their cells. Looking forward, women pass on their mitochondrial DNA to all their children while men do not. But looking backward, you got your mitochondria from your mother. She got it from her mother, who got it from hers, who got it from hers and so on back into the remote past. This brings us to the logical but nonetheless aston-ishing conclusion that, at any time in the past you care to choose, be it a hundred, a thousand, ten thousand or a hundred thousand years ago, there was only one woman alive at the time who passed her mitochondria down to you. To temper scientific rigour with a touch of romanticism for a moment, I often wonder what she was like, my maternal ancestor, whose mitochondria I am using each time I breathe. Where was she in 1066, or during the days of the Roman Empire or during the depths of the last Ice Age? All I can be sure of is that she survived and had a daughter. There is a mirror image of this maternal inheritance on the male side which could, in principle, and increasingly in practice, be traced using the Y-chromosome as it is passed from father to son.[8]

These are much more direct connections to an individual than through the nuclear genes, which are inherited equally from both parents. For nuclear genes, we double the number of ancestors with each generation. Four grandparents, eight great-grandparents and so on. By the time we have gone back only twenty generations, about four hundred years ago, there are one million potential ancestors contributing bits of your nuclear DNA. In practice a lot of them will be the same people, but among them will be that one woman, the maternal ancestor whose mtDNA has been handed down to you. We are all mongrels, just a sample of the many different alleles[2] that happen to be in the population at the time. That's what makes the often-asked question—'Where do I come from?'—in one sense unanswerable. We are all a mixture, with every gene, every base, having a different pedigree. It reminds me of a famous exam question which involves a proof (not particularly difficult) that each time we breathe we inhale a few molecules of Caesar's dying breath. As too with genes. Given enough time for diffusion, we all of us have genes from Africa, from Asia, from Australia, maybe even genes from Neanderthals and other extinct archaic humans. The answer to the question is—'From everywhere'. Only the proportions differ.

Turning back, hurriedly, to the comparative simplicity of mito-chondrial DNA, another logical inevitability is that any two individuals can connect themselves through a maternal lineage. At some point there will be a common maternal ancestor—who would not only have survived but also had *two* daughters. The only difference between any pair of individuals is the length of time since that woman lived. That is where another useful feature of mtDNA comes in. When cells divide, DNA is copied. This is a remarkably efficient process—as it has to be—but very occasionally errors are introduced. These are mutations. In nuclear genes, because the copies are being constantly scanned for errors, the mutation rate is very low but in mitochondria, which lack a good error checking system, the rate is much higher. If you compare the mtDNA sequence between two individuals, any differences between them must have accumulated since the common maternal ancestor. You can if you know the mutation rate, estimate how long ago she lived.

Though mutations occur in many forms, we use the very straightforward substitution of one base by another as the basis for our comparisons. For practical purposes we scan about 400 bases in what is known as the mitochondrial control region which, because it doesn't code for proteins, retains mutations more easily. The DNA language has a four-letter alphabet. Comparing two sequences is exactly analogous to comparing two different spellings of a 400-letter word written with this simple alphabet. The rate of change for this region between two individuals works out at about 1 substitution in 10 000 years. If two individuals differ at only one position in the control region this means that, on average, their common maternal ancestor lived about 10 000 years ago. Differences at two positions push that time back to about 20 000 years and so on. As with individuals, so with populations. Theoretically, everybody can be connected, and the patterns these connections make and the times involved are the basis of this new type of population genetics.

Could it be that simple? When we were thinking of applying this logic to humans, I remembered having read as a boy that all Golden Hamsters in the world were derived from a single pregnant female who was dug up from her burrow in the Syrian desert during the 1930s. I always thought this sounded a highly improbable story but, if it were true, all Golden Hamsters should have the same mtDNA, modified only by mutations which had arisen subsequently. Over one summer we contacted Syrian Hamster (their correct name) societies around the world, and having worked out a way of recovering hamster mtDNA from the relatively innocuous droppings, had no difficulty in getting samples. When we sequenced the control region we found every hamster, and we tested about seventy, had exactly the same sequence. So the results confirmed the story and, more importantly, we felt confident that the control region was sufficiently stable to make a human project feasible.

5.3   POLYNESIA

The story now moves to a remote location—Rarotonga—largest of the Cook Islands in the South Pacific (Fig. 5.4). I stopped there for a few days while flying from the United States to Australia, hired a motorbike, fell off, broke my shoulder, and was confined to the island for several weeks. As luck would have it, there is a thriving library on the island with a very well stocked section on Pacific history and anthropology. There I read about the competing theories on the origins of the Polynesians. No-one who has ever flown over the vast empty ocean for hour after hour can help but wonder at how the original settlers had managed to cross these vast stretches of open water without power or navigational instruments. All the islands in the Pacific, no matter how remote, had been reached before the Europeans arrived, beginning in the 16th century. The Englishman, James Cook, by no means the first but certainly the most widely travelled of Pacific explorers, noticed that wherever he went the languages were mutually intelligible. This common language, and their similar physical appearance, hinted at a common origin for the islanders across this vast ocean.

But where had they come from? There are two competing theories. The majority view, based on linguistics, on a ceramic culture called Lapita, on crops and domesticated animals, strongly suggested that the majority of Polynesians had come from somewhere in South-east Asia, arriving in Samoa and Tonga about 3000 years ago. From here they spread to the Marquesas Islands, thence to Hawaii in the North, over to Easter Island in the East, and lastly down to New Zealand, which they reached about 1000 years ago. In complete contrast, the Norwegian anthropologist Thor Heyerdahl put forward the view, based on some word connections, oral mythology, and, most important, the widespread

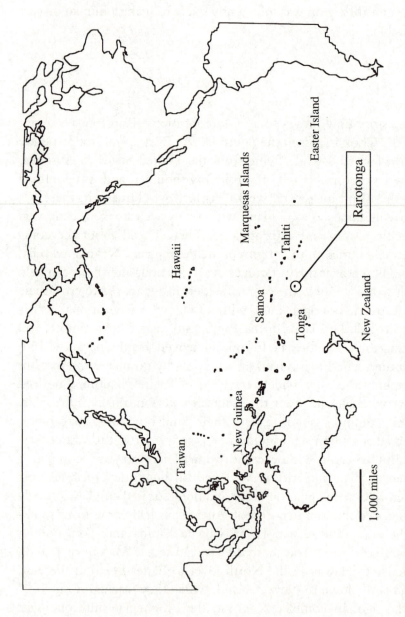

Fig. 5.4 The Pacific, showing islands mentioned in the text.

cultivation of *kumara*, the sweet potato of undisputed Andean origin, that Polynesians came largely from North and South America. To demonstrate that this could be done he famously sailed, or rather drifted, his raft *Kon-Tiki* from the coast of Peru over to the Tuamotu islands east of Tahiti. The two theories reflect on the skill and motivation of the first explorers. The winds and currents in the Pacific are almost always moving from East to West. A colonisation from South-east Asia would have meant deliberate and skilful navigation against wind and current while an American origin may have meant, as some anthropologists disingenuously proposed earlier this century, that Polynesians just drifted onto the islands accidentally while out fishing.

Despite a great deal of data from blood groups and other classical markers over two decades, the question had not been decisively resolved so I wondered whether mitochondria could shed any new light. Collecting twenty outdated blood samples from the hospital, I virtually smuggled them back to Britain and sequenced the mitochondrial control region. The results were astonishing. Sixteen were exactly the same, three had one change and one was completely different, with twelve sequence changes from the common type (Fig. 5.5). The circles represent particular control region sequences or *haplotypes*, and the size of each circle is proportional to the number of people that share the haplotype. In the Cook Island sample one haplotype is clearly predominant. There is a less common but closely related haplotype one mutation away, and a third, completely different haplotype. I originally thought this must have come from a visitor to the island because these were just random samples from the hospital outpatient department. The results nonetheless warranted a return visit the following year. On the way I called on Rebecca Cann in Hawaii. She and her student Koji Lum had been sequencing mitochondrial control regions from native Hawaiians and I vividly remember the excitement we all felt when we compared results and discovered that the unusual haplotype I had found in the Cooks exactly matched one of their native Hawaiians. So it was not a visitor but a genuine Polynesian. I collected from other places in Polynesia, and with the help of samples from colleagues and published results from the Americas, we traced the Polynesian haplotypes.[9] This confirmed the existence of the two very distinct Polynesian groups throughout the region. The common group, comprising about 95% of the sample, came from South-east Asia with the closest links to aboriginal Taiwanese—which is also the probable origin of the Lapita ceramic culture. The rarer group, at about 4% of the sample, did not come from South-east Asia but from New Guinea. There were about 1% of mixed types that included two, out of a thousand Polynesians, which matched published Chilean sequences.

It is very clear what has happened. The majority of Polynesian mitochondrial DNA had come from South-east Asia, picking up perhaps only one woman from New Guinea on the way. This is a crystal-clear result. There is no evidence of a large scale colonisation from North or South America. Heyerdahl was wrong. Perhaps the two 'Chilean' sequences were brought in by two women who had boarded Polynesian ships on their way back from South America during voyages of the sort that introduced the sweet potato. Might I here insert the usual caveat. Mitochondria only trace the history of women and we cannot formally rule out a coincident movement of women from South-east Asia and men from North and South America who happened to meet somewhere in the middle of the Pacific!

The major Polynesian haplotype is very easy to recognise and it has never been found in South America. The Polynesians seemed to dislike settling on anything but virgin territory. The only other place they have been found is Madagascar. It has been known for

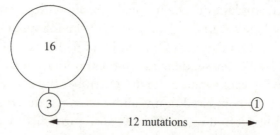

Fig. 5.5 The first results from 20 Rarotongans. Circles denote separate haplotypes defined by their mitochondrial control region sequences and the number of individuals with that sequence is shown inside the circle.

some time, from linguistics, that the island had strong links with Indonesia and the mitochondrial evidence has been used to support an Indonesian origin by slow progression around the coasts of India, Arabia and Africa. I much prefer to think that Madagascar was colonised directly from Polynesia itself. This would have involved going to the south of Australia and across the Indian Ocean—an immense voyage but one which, having learned something of their amazing feats in the Pacific, I can quite easily believe they were up to.

The result in Polynesia was a demonstration of the power of mitochondria to produce a clear answer when decades of other investigations, although they had got the general idea, had never reached a convincing conclusion. Blood groups could never have delivered such a clear verdict anyway because the same blood groups are shared by native Americans and Polynesians owing to their common Asian ancestry. Neither would we have been able to detect the small but nonetheless real New Guinea contribution from the genetic distance tree (Fig. 5.6). This shows a very deep separation between Polynesia and New Guinea and completely misses the connection. I think it's fair to say that mitochondria are not popular among people who draw genetic distance trees because the results don't fit the trees. When I am challenged about

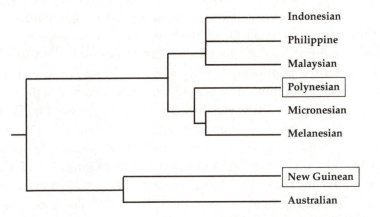

Fig. 5.6 Part of the genetic distance tree, redrawn from Fig. 5.1, showing a very deep split between New Guinea and Polynesia. In contrast, mtDNA shows that mitochondrial lineages from New Guinea are found in Polynesia.

this, as I often have been, the challenge is along the lines 'Well, they don't fit so there must be something wrong with mitochondria and how you're analysing it.' I think the real answer is that they don't fit because there is something wrong with the trees.

5.4   EUROPE

Let me now turn to Europe, which is—not surprisingly—a much more difficult proposition than Polynesia. Leaving aside Boxgrove Man and other early examples of archaic humans, Europe was inhabited by the Neanderthals for at least 200 000 years before being replaced by, or multi-regionalists would say transforming into, anatomically modern humans from about 40 000 years ago. The most recent Neanderthal fossils are found in Southern Spain and are just under 30 000 years old—so there was a period of 10 000 years during which both Neanderthals and modern humans shared the continent. About 10 000 years ago, farming began to spread from the Middle East, reaching South-east Europe around 8500 years ago and spreading slowly and unevenly until it reached Britain two thousand years later. There is a gradation, a cline of 'classical' gene frequencies from the Middle East to North-west Europe. The fit between this and the spread of agriculture led Ammerman and Cavalli-Sforza to formulate what they termed the 'demic diffusion model' for the Neolithic transition.[10] Agriculture allows the population density to increase and pressure at the margins forces a diffusion of people outwards following a process, a 'wave of advance', first elucidated as a statistical model by R.A. Fisher. The implied consequence is that the Mesolithic population, the hunter-gatherers, were absorbed or completely overwhelmed by this expansion. In fact there aren't figures put on the quantitative impact of the expansion[3] but I think it is certainly fair to say that most people interpret their demic diffusion model as a massive and overwhelming incursion, more or less swamping the indigenous hunter-gatherers. For instance, Colin Renfrew, commenting on the Basques, writes: 'the high rhesus-negative gene frequency and the (non-Indo-European) Basque language would both be seen as relics of the early, pre-Indo-European population of Europe *that was effectively swamped, genetically and linguistically, in the farming dispersal process.*' (my italics)[11] leaving us in no doubt of the importance of this process for the composition of the modern European gene pool.

So what do the mitochondria tell us? Do they support the majority view, as was the case in Polynesia, or not? We collected together, from our own sampling expeditions and, as before, with the help of colleagues and some published data, the control region sequences of just under 900 people from Europe and the Middle East.[12] When these are put into a network, the picture, not surprisingly, is much more complex than in Polynesia (Fig. 5.7). But there are some similarities. There is a very common haplotype, the CRS or Cambridge Reference Sequence, so called because it was the first mitochondrial sequence to be worked out, in 1981, in Cambridge. Radiating outwards from the centre are some closely related 'satellite' haplotypes that are one- or two-step derivatives

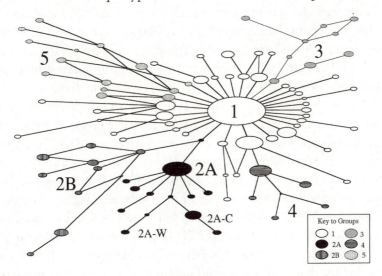

Fig. 5.7 A network of the commoner European mitochondrial haplotypes. Ovals are different sequence haplotypes with areas proportional to their frequencies in the sample. Their most likely evolutionary relationships are denoted by connecting lines and when there is uncertainty two or more routes may be given. The lineage groups have different shading denoted by the key.

of the central type. In fact we can see this process beginning in Polynesia, where about 5% of the haplotypes have gained an extra mutation, probably since they arrived in Polynesia. If you left Polynesia alone for 25 000 years or so and repeated the analysis, having allowed further mutations to accumulate, then you can show you get something which looks like the European central haplotype and its satellites.

But there are lots of other branches in the network. Because we couldn't have interpreted what was going in Polynesia without dividing the population up into different mitochondrial groups, we attempted to do the same thing in Europe. It's clearly nothing like as easy—nonetheless we've tried it and have divided the population into six groups (1, 2A, 2B, 3, 4 and 5) and looked to see whether there are any connections with other parts of the World which might indicate founder populations. Starting with the most diverse, and therefore perhaps the oldest, group: Group 3. It is found mainly at the western margins of Europe but has its clearest mitochondrial relatives in Africa. The group is also found in the Middle East and we interpret it as the echo of the first migration of anatomically modern humans out of Africa. We have found the connections, mainly in the Middle East, to the immediate ancestor of the CRS from which most of the other European groups are ultimately derived. The group whose derivation from the 'old' CRS is the least clear is Group 5. It is rather an untidy group, quite diverse and much commoner in Northern Europe than in the South. Group 4 we find throughout Europe, and although the diagram does not show this because it is constructed from only the commoner types, the central Group 4 haplotype is surrounded by a ring of 'satellites' indicative of a population expansion some time in the past. This is a feature shared with the CRS and its satellites, which we call group 1. Group 2B is widespread and diverse but Group 2A is quite different. There is considerably less diversity in Group 2A than any of the others and it has a fascinating geographical distribution. It divides into three smaller groups, two of which are geographically restricted, as we will see in a moment.

How do the features of this network fit with current theories? First, there is no sign of surviving Neanderthal mtDNA in the network. If we take a date to the last common ancestor of Neanderthals and anatomically modern humans (AMH) of 250 000 years, then there would be about 25 sequence differences between a typical Neanderthal and AMH sequence.[4] We would see two well separated clusters if the ancestry were mixed, even further away from each other than the Papuan and South-east Asian groups seen in Polynesia. But we don't. Although there are different groups in Europe, they are not separated by anything like that mutational distance. Neither are there any individual haplotypes that are so far adrift from the rest as to be credibly Neanderthal. So we have not detected any signs of interbreeding between Neanderthal females and AMH males. At least not yet. Mitochondrial DNA does not support the multi-regional view for Europe, and agrees with the complete replacement of Neanderthals and their eventual extinction.

How does the mtDNA picture fit with the demic diffusion model, or at least its popular incarnation, which sees the older, Palaeo- and Mesolithic populations being swamped by Middle Eastern farmers? First of all, the divergence dates for the different groups do look rather too old. By divergence time, I mean the time it would have taken for all the observed variation to have accumulated from the common ancestor. They range from about 50 000 years for group 3, 35 000 years for groups 2B and 5 to around about 25 000 years for group 1 and rather less for group 4. Only the sub-groups of group 2A are in the right range of about 10 000 years. These dates are only approximate and rely on a good estimate of the mutation rate. Having said that, if you repeat the operation in Polynesia, you do get a good agreement between the accumulated diversity and the archaeological dates for colonisation using the best estimate for the mutation rate—so it can't be all that far off.

However divergence dates and settlement dates are not necessarily going to be the same. There would only be a good correspondence between the two if all the diversity had arisen *after* the settlement, as is the case in Polynesia once the two lineage groups had been identified as having different origins. Diversity that had accumulated *before* the settlement will push the divergence date further back, beyond the settlement date by a factor which depended on the diversity of the incoming population. Although we don't yet completely understand how this factor behaves, for example what influence population size fluctuation might have on the outcome, one should expect to find a good agreement between the mitochondrial structure of the founding and settled populations for the settlement and divergence dates to be widely at odds. In fact the mtDNA pattern in the Middle East is really very different to Europe. In the Middle East, group 1, which embraces over half the European haplotypes, is rare—and the CRS itself absent—and the main lineage group is 2A. The other groups are present but very few of the haplotypes match those found in Europe. This is not what one would have expected from the 'massive demic diffusion' model—we should have seen much more similarity between the populations than we do. One might argue that we are taking the wrong founding population. If we compare Europe to Anatolia the structures are certainly more alike but there are still precious few haplotype matches. And besides, the cline of gene frequencies that forms the basis of the demic diffusion model does not show a large distinction between Anatolia and the Middle East.

Suppose we are being badly misled by the dates and that the mitochondrial pattern in Europe is basically that of the Neolithic settlers. In that case, Group 1 is so common it would certainly have to be Neolithic. Apart from the Saami from northern Finland, there is only one population that is widely thought to be a relic of the pre-Neolithic population: the Basques. Agriculture came very late to the Basque country of northern Spain and the Basques are also distinguished as speaking the only extant non-Indo-European language in western Europe.[5] If Group 1 had been a major component of the agricultural demic diffusion then we would expect low frequencies in the Basques. But, in fact, the Basques have the highest Group 1 frequency of all—about 77% compared to the European average of 60%. If they are a relic population then Group 1 was present in Europe before the Neolithic and was not introduced by the farmers.

If Group 1 were Palaeolithic, are there any good candidates for a mitochondrial cluster associated with the Neolithic farmers. Well, there is one, Group 2A, and particularly the geographically restricted sub-groups we call 2A-Central (2A-C) and 2A-West (2A-W). Altogether they make up 12% of the European mtDNA lineages. The ancestral sequences are found in the Middle East and Anatolia and the haplotypes are very similar throughout Europe, suggesting they haven't been there all that long. They are also virtually absent from the Basques. Astonishingly, there is a very clear division between the distribution of the two sub-clusters, with 2A-C being found in Central and Northern Europe and 2A-W found along the Mediterranean and Atlantic coasts. No other lineage groups are so constrained in their distribution. We have proposed that these are the signals of the Neolithic settlement, following, as they do, the two archaeologically defined cultures— one along the Mediterranean and up the Atlantic coast to Britain, associated with the Impressed Ware ceramic culture, and a second along the valleys of the Danube and Elbe associated with the *Lineanbandekeramik* (LBK) ceramics.

There is one more piece of evidence worth mentioning. We have extracted DNA from a 13 000 year tooth found in Gough's Cave in Somerset. It is slap bang in the middle of Group 1. This is direct evidence that this common haplotype was already present in Britain at least 6000 years before the arrival of farming. To us the simplest explanation for the observed mtDNA patterns remains that the mitochondrial ancestors of most modern Europeans arrived during the Upper Palaeolithic and not the Neolithic. In overall terms, this means that, over time, the indigenous population must have learned to farm, perhaps from pioneers defined by Group 2A lineages. Although Colin Renfrew has firmly linked his ideas about the Indo-European language family to the demic diffusion model, I don't see a problem with the spread of farming, by a substantial but not overwhelming group of Neolithic incomers, with the spread of language being a coincident process.

So to go back to where we started: the gene–language diagram. As I have endeavoured to explain, I think the left hand side—the gene side—is misleading if it is interpreted as an evolutionary history. Human populations, apart from exceptional cases, just don't split and remain ever after separated. Their gene pools are mixtures. Individuals are just samples of the gene pool, temporary combinations of alleles for one generation. It follows quite automatically that the larger the number of loci you study, the nearer any individual will get to the population average *defined by the same sample*. The argument is circular. It is bound to happen.

Small wonder that mtDNA doesn't fit comfortably with the world according to population trees. It is the first locus where a reasonable stab can be made at the evolutionary relationship between alleles. We will have to wait and see what happens with other loci with which you can do this—something that is now happening with the Y-chromosome and the β-globin locus[6] on chromosome 11. My prediction is that none of them will 'fit' and we will have to redefine, or even abandon, the notion of human populations as a useful taxonomic unit. I have referred already to the caustic feel of contemporary correspondence on the subject. Arthur Mourant, the doyen of practical population genetics and the great compiler of distribution maps, said much the same thing fifty years ago in the milder and far more elegant style of the time: 'Rather does the study of blood show a heterogeneity in the proudest of nations and supports the view that the races of the present day are but temporary integrations in the constant process of mutation, selection and mixing that marks the history of every living species'.[13]

NOTES

1. Genetic drift is the process whereby gene frequencies change between generations purely by chance. It is due to the random sampling of parental alleles into the germ cells. The effect of drift to alter allele frequencies is much more stronger in small populations than in large ones.
2. Alleles are the different versions of a gene that are found in a population.
3. Since this lecture was delivered, Cavalli-Sforza and Minch have put a figure of 26% on the overall contribution of the Neolithic incomers.[14]

4. We now know one Neanderthal sequence. It has 27 control region differences for the CRS.[15]

5. Other than Finnish and Estonian.

6. Locus (pl. loci) is a position on the genome. It is for practical purposes synonymous with 'gene' but, as a more general term, also refers to stretches that are not genes because they have no function or, as here, that contain several genes.

6

Ancient DNA

Svante Pääbo

The vast majority of all species that have ever existed on earth are now extinct and artefacts of only a small minority survive, either as museum specimens or as fossils. Nonetheless, if we could retrieve genetic information from these remnants it might add a great deal to our knowledge about them, about their relationships to each other and to living populations. Although indirect genetic information has been retrieved from fossils for many years in the form of blood-grouping and immunological cross-reactions of recovered proteins, in addition to straight morphology which has a strong genetic component, it is the retrieval of DNA, the primary genetic material, that I want to address in this paper. In doing so I will begin by a more technical discussion, because there are still some technical problems with retrieving ancient DNA, especially from humans. Then I will go further back in time to work done on some extinct species, and end by discussing the possibility of going way back in time to fossils that may be millions of years old.

It all began in the early 1980s, when we and others starting looking at the microscopic appearance of ancient Egyptian mummies and comparing them with a contemporary human muscle that had been air-dried to mimic the mummification process. In the artificial sample we could see nicely preserved muscle fibres with the cross striations[1] intact and cell nuclei clearly visible. If we compared this to a muscle sample from 2500-year-old Egyptians prepared in the same way, the picture was usually very disappointing. In these ancient samples we could barely see the muscle fibres and there was no sign of any intracellular structure. Very occasionally, however, we found a sample that looked more encouraging. For example, we looked at a sample from the left leg of the mummy of a child from Egypt that was around 2700 years old. In the deeper levels of the skin we did see what might have been cell nuclei. When we stained these sections with a dye that will bind to DNA and fluoresces when illuminated by ultraviolet light, we could see that these structures did light up. So here the DNA seemed to have been preserved. Could it be extracted so that the crucial information, its sequence of bases, could be read?

The first step is always to grind up the tissue and purify the DNA away from the other molecules—proteins, fats, and so on—that are present. Having done that, the next step is to get an idea of the lengths of the DNA fragments. This we do by a process called electrophoresis, where a few drops of the DNA extract are put into a small well formed in a gel. When a small direct current is passed though the gel, the DNA moves towards the positive terminal. The smaller fragments find it easier to get through the pores in the gel and so migrate faster than the larger fragments. After an hour or so the gel is stained with the same kind of dye that was used to see the DNA in the muscle sections. You can then see the DNA fragments and, by their final positions in the gel, estimate their lengths. In the ancient extracts the DNA is only ever about 100–200 base pairs[2] in length. This is in dramatic contrast to modern DNA prepared in the same way where the fragments are much larger: from 10 000–100 000 base pairs (bp) long. So the ancient DNA is degraded. It doesn't seem to matter how old the specimen is. We saw the same short lengths of DNA from a four-year-old piece of pork that we happened to have in the laboratory, in a 130-year-old extinct marsupial wolf, in human mummies from Egypt and South America that are a few thousand years old, and also in a 12 000-year-old extinct ground sloth. And they are all degraded to approximately the same size. It looks as if this reduction in size is probably something that happens rapidly after death. But there are also other types of damage that accumulate over a long time. If you analyse the individual DNA bases that you extract from ancient remains by mass spectroscopy, which we have done recently, you find a lot of different modified base structures, particularly those that have been modified by oxidation. When there are such base structures in the DNA, they block the activity of the enzyme we use to copy DNA in the test tube.[3] When we look at the amount of such modified bases in a set of samples and compare that with our ability to amplify DNA from it, we find that, however many attempts we make, we fail when there are large amounts of oxidised bases. So oxidative damage to the DNA is a major factor in preventing the recovery of DNA by the polymerase chain reaction.

You can also look directly at the DNA using an electron microscope. As expected, most DNA molecules are very short but you do come across occasional longer molecules. You don't know if these are really ancient or whether they come from fungi or bacteria that have grown, or perhaps are still are growing, in the remains long after organism that died. You will also see structures where DNA molecules have become cross-linked to each other. So there is always a large amount of damage in ancient DNA. We are used to amplifying sequences from modern DNA that are several thousands of base pairs long, whereas in the ancient samples we can only do short pieces. So if we want to determine a longer sequence, what we have to do is combine many overlapping short amplifications. To take an example, when we analysed bones from extinct ground sloths from southern Chile, we could quite easily amplify fragments 200 bp long but as we increased the length to 287 and 394 bp it became progressively worse until finally, at 540 bp, it didn't work at all. When we parcel together the short sequences to make one long sequence and do a comparative analysis of this mitochondrial sequence with those of living sloths, we see that the extinct sloth, which is around 12 000 years old, clearly falls together with the two living tree sloths from middle America. So we have a lot of confidence that these sequences are indeed ancient sloth sequences. However, I should also point out that not all remains work. Of 35 different sloth remains that we have tried only two work, and they were the ones from southern Chile.

We got very enthusiastic and tried to amplify nuclear[4] DNA sequences from the same samples. When we tried ribosomal

DNA, which is still repeated in many copies, we found we could amplify something but we also observed that the longer the fragment the better the amplification worked. When we analysed the sequence we found there were many differences from both modern sloths and humans but we found it was identical to a fungus. This of course doesn't mean that the sloths were related to fungi, but simply shows that we have a contamination problem.

A big problem in much of this work on ancient DNA concerns authenticity. How sure are we that the sequences we retrieve are really from the organism that we want them to be from? The biggest problem is actually contamination with contemporary DNA. This is because the polymerase chain reaction is incredibly sensitive and can amplify even from single DNA molecules that are present. Without this sensitivity it would be impossible to get anything from ancient remains. But this astonishing sensitivity also means that it is extremely easy to amplify any molecules of modern DNA that get into your extracts. Since all the ancient molecules are modified to some extent and don't amplify very well, any undamaged, modern DNA will be amplified preferentially and you may well end up with just modern sequences. So, to avoid contamination, you have your collaborators or your students dress up in interesting clothes and do the work in specially dedicated laboratories. And still you will have problems. For example, when we worked on moas, which are extinct, large flightless birds from New Zealand, we used specific moa primers in the amplification reaction on some bone and found no specific amplification. However, when we used primers specific for human sequences, one of the extracts gave a product which we sequenced and, sure enough, we obtained a nice human sequence. So it is clear that this sequence comes from an archaeologist, a museum curator, or even from one of us in the laboratory.

The problem here, of course, is that if it had been human bones we were studying, it wouldn't have been so easy for us to identify this as a contamination. So you have to fulfil a number of criteria before you believe in your results. There are various controls to make sure the reagents are not contaminated and you must make sure that multiple extracts from the specimen will give the same sequences and so on. Where you can, you need to be able to make sense of the sequence so that a sloth sequence should be sloth-like, and so on. However the ultimate criteria, I think, for all of this work is independent reproducibility. It is important to show that other laboratories can repeat your work from the same remains. This means, when it's important, removing several samples from one specimen and sending material to other laboratories as we and Bryan Sykes have done in some cases. However, even when you have satisfied all these criteria, it ends up being quite difficult. For example, in a study in which we analysed 110 mummy skeletons from ancient Egypt, all of which were morphologically very well preserved, only two yielded results which are reproducible and that we believe in. So in many circumstances, this is very, very difficult.

There have been only a few studies where real insights about human population history have come out of studying ancient DNA. As an example of one of these, I would like to refer to work done by Anne Stone at Penn State University in Mark Stoneking's laboratory. It involves remains from the Americas where, as you know, humans arrived from Asia maybe 13–14000, or perhaps even 30000, years ago. If you study mitochondrial DNA in present day Native Americans and compare it to Asian populations, what you find is that most mitochondrial lineages in America tend to fall into four groups, each of which has clear antecedents in Asia. You also find much more diversity in Asia than in the Americas. There is a reduction in diversity in Native Americans relative to many other groups in the world, and the common belief is that this is due to the colonisation by a relatively small number of individuals.

There is another formal possibility, however. The reduction in population size of Native Americans that followed contact with Europeans could have led to this reduction in diversity. It is difficult to know how many Native Americans there were when the Europeans first showed up, but it may well have been be in the few millions. This number was drastically reduced due to epidemics and genocide in different forms, and is only now beginning to increase again. So to look into this, Anne Stone studied a 600-

year-old archaeological site in Illinois, Norris Farms, south-west of Chicago. Over 200 skeletons have been excavated from Norris Farms and many are extremely well preserved. Interestingly, it seems as if violence was a big problem then, just as it is now, because more than 30% of the male skeletons died due to some sort of trauma. Anne identified four mutations in mitochondrial DNA that are typical of the four groups of related lineages. These were three point mutations[5] and one deletion of nine base pairs. This allowed her to do the very short amplifications that work well for ancient DNA and then test the extracts for these characteristic differences. This was straightforward because she could detect the 9 bp deletion just from the size of the fragment and could spot the point mutations because they occurred at restriction sites.[6] So she could then look for the size of the fragments simply for the deletion, or digest the product with a restriction enzyme to look for these point mutations. She was able to reproducibly analyse DNA from 50 skeletons from the cemetery and compare the results with present day Native American groups. What she discovered is that there is no statistical difference in the diversity between the ancient and modern populations showing, not very surprisingly, that this reduction in diversity probably owes its origin to the initial colonisation of the Americas during the glaciation rather than the reduction in size after European contact.

However, I would like to add a note of caution. This comes from work that we have done on just three mummies, from the Hohokam culture in south-western USA, that are around 600 years old. They are morphologically extremely well preserved. What we did was to estimate the number of DNA molecules in the extracts. The simplest way to do this is to add into the bone extract known amounts of the target sequence, say a segment of mitochondrial DNA, into which you have engineered a short insertion of 20 bp or so. You can then amplify the different mixtures, look at the results on a gel, and compare the brightness of the band coming from the ancient DNA and that from the artificial template which is 20 bp longer. We might add a range of artificial template concentrations, say from 10 000 molecules down, in sensible steps, to 1 molecule. With 10 000 molecules the band from the artificial template is the only one visible but, as this is reduced there comes a point when the bands from the ancient DNA matches it for brightness. By assuming that the number of molecules of each template is about the same at the balance point we can estimate how many molecules of ancient DNA were in the original extract. When we did this with the three Hohokam mummies extracts from one contained about 1000 molecules while from another, apparently equally well preserved, there were only single molecules of DNA present by this test.[2] We then took the analysis one stage further by cloning the amplified DNA into bacteria[7] so that we could sequence single molecules. When we amplified across a small region of mitochondrial DNA that shows a lot of variation in modern humans and cloned them we found that, in the mummy with 1000 molecules of DNA in the original extract, all the sequences contained the same set of seven variants. There were also some additional differences in individual molecules which we interpret as being due to errors during the amplification, probably induced to some extent by damage in the template. When we make a second extract and again do the same analysis, we find the same seven positions differing from the reference sequence.[8] So we believe that this is a true sequence. What happened when we looked at the clones coming from the extracts with only a single DNA molecule or so? When we made the first extract we were quite happy with our results. We have five positions where all clones carry the same variants. Then we make a second extract and we find three positions where all clones carry the same substitution, but they are at different positions than in the first extract. In addition, we find positions which are ambiguous, where some clones carry substitutions and others don't. We make a third extract. Again we have different sequences from the other ones. And then a fourth extract and again we have different sequences. So here we simply can't determine the original sequence. I think it shows how useful it is to try to reproduce your results because you find, in practice, that you cannot reproduce them very often. It also shows, actually, that it is very useful to clone your products and look at many clones to discover problems like this. So I think that quantitation and cloning are very useful things to do. The reality is that the contribution of ancient DNA to studies on human history has been very limited so far, and these technical problems are the reason.

So I will leave human remains and step backwards in time because most of the cases where we have been able contribute something to real understanding come from the study of extinct species. I will just show one example and that involves the large flightless birds that used to live in New Zealand: the moas. Some of them stood two or three metres tall and weighed several hundred kilograms. The work I will present has actually been performed by Alan Cooper who now works in Oxford. The moas have been very important in New Zealand. Since there were no land living mammals, except bats, they had hardly any enemies except for a very large extinct eagle—that is, until humans arrived. Moas soon disappeared following the Maori settlements from 1000 AD when, being an easy source of prime meat, they were hunted to extinction. Moas have influenced both the flora and the fauna rather dramatically. There are several species of New Zealand plants, for example, that have a strange way of growing— verication—where the branches cross over one another and the leaves are protected by the branches, which people believe is a protection against grazing by moas. Moas are part of a larger group of birds, the ratites, that all exist on the southern land masses. There is the ostrich in Africa, the rhea in South America, the emu and cassowary in Australia and New Guinea, and, of course, the kiwi in New Zealand. These birds are all flightless, and are all related to each other and to the tinamous of South America, which have a limited capacity for flight. These related species came to live on the different continents in the southern hemisphere because of a common ancestor, itself flightless, that lived on Gondwanaland.[9] When plate tectonics led to the formation of the different continents, the populations became isolated from one another and began to evolve independently. There was a land bridge from New Zealand to Antarctica and Australia until something like 80 million years ago. The idea was that there was an ancestor of both kiwis and moas that became isolated on New Zealand and subsequently evolved into the different kiwi and moa species. This would then mean that moas and kiwis would be each other's closest relatives. To test this, Alan Cooper assembled remains from many of these moa species.[3] They generally came from cave sites where in some instances you can find bone and even soft tissues from these moas that are a few thousand years old. He found that the remains from cave sites were so well preserved that he could amplify segments of 400 bp from bone, but not from skin. Alan compared the mitochondrial sequences from four moa species, the three kiwi species, and the living flightless birds on the other continents and then constructed the phylogenetic tree (Fig. 6.1). All the moas are closely related to each other. The kiwis are also related to each other but they are not at all related to the moas. The big surprise then is that the moas are an early divergence among these groups of birds, whereas the kiwis are more closely related to the emus and cassowaries of Australia and New Guinea, and even to the ostrich from Africa, than they are to the moas. So whereas it is easy to imagine that the moas became isolated on New Zealand 80 million years ago, the question becomes: how did the kiwis get there later, presumably from Australia, if they are flightless? So the idea emerged that perhaps flight had been lost several times independently and that the ancestor of all these birds was not flightless after all and that the first kiwi ancestors flew to New Zealand.

That was, of course, a very revolutionary idea. Not everybody agrees with it, but some people have begun to rethink the evolution of this group of birds. In the meantime, Alan Cooper has

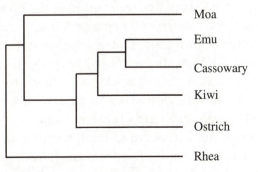

Fig. 6.1 Molecular phylogeny of extant and extinct ratites reconstructed from mitochondrial DNA. (Redrawn from Cooper *et al.*[3])

since reformulated this a little bit. There has now been work on other groups of species in Australia and New Zealand, for example the Southern Beech. This shows that the closest relative of the beech in New Zealand actually exists in Australia and New Caledonia[10] and that the split seems to go back something like 50 million years. And for another group of birds, the bower birds, though they are not found in New Zealand, the situation is the same. So maybe there has been some opportunity for several species to move from Australia to New Zealand, until 40–50 million years ago, via islands such as New Caledonia.

There has recently been a rather dramatic demonstration that New Zealand hasn't been as isolated as was once supposed. It had always been thought that there were no dinosaurs on New Zealand because it had been isolated for so long; that was, until a lady went out and found a dinosaur fossil in her back yard in the outskirts of Auckland. And that then brings us to the last part, going even further back in time, and to asking the question: can we now think of retrieving DNA from remains such as these? To answer questions about how species evolved over millions of years would be extremely interesting to evolutionary biologists, of course.

So maybe it would be possible—and so it would seem from the paper that appeared three years ago in *Science*,[4] where DNA sequences were presented from 80-million-year-old bones that had been found in Utah and were clearly of dinosaur origin. The paper presented nine different mitochondrial sequences extracted from this bone that looked like really nothing else on earth that we know of. However, when we compared these sequences with those from other animal groups, it was rather surprising because sequences from this bone did not really group together with birds, as you might expect dinosaurs to do, but tended to cluster with humans. But they were clearly not human because some of the sequences were 30% different from contemporary humans. So, Hans Zischler, in our lab, began to think about what these things might be that look quite like human mitochondrial DNA sequences but yet are clearly not identical to them. What he was thinking of were integrations into the nuclear genome of pieces of mitochondrial DNA. These would be pieces of DNA that had broken off from the mitochondria and been integrated into the nuclear chromosomes of a remote ancestor. Once there they would evolve in a different fashion and might accumulate mutations that make them look quite different from the mitochondrial copies of modern humans. In order to test that possibility, we needed modern human DNA that was absolutely free of mitochondrial DNA. And that's not totally easy to do. First, you have to have a least some males in your lab that are willing to co-operate. Sperm heads contain only nuclear DNA because all the mitochondria in the sperm are located in the mid-piece. Using the same primers that they used in Utah on highly purified sperm-head DNA, we did indeed amplify a band. When we sequenced it and compared it with the data from the *Science* paper we found it clustered very nicely with the nine 'dinosaur' sequences from Utah.[5] There are different formal possibilities about what could have happened in this case. One of them is that we have a contamination in our laboratory of dinosaur DNA—which we do not think is likely. One could also think about hybridisation in the late-Cretaceous, perhaps between mammalian ancestors and dinosaurs, which is also interesting but highly unlikely. And in the third case, there could be a contamination of human DNA in the fossils in Utah.

Although it is easy to be funny about this, I think it shows really how difficult it is to do this type of work. That even if you do all the controls you can think of, there may be some source of contamination, such as nuclear insertions in this case, which is not easy to think about. So it would be very useful to have some other means of analysing the remains to see whether they are likely to contain DNA in a form that would be amplifiable. It would be especially good to have a screening method which would allow you to quickly analyse many remains to find the few where conditions have been such that macromolecules like DNA might survive. To do just this we have begun to look not at DNA, but at amino acids which are the building blocks that make up proteins. As you may know, amino acids that are incorporated into proteins are introduced in the L-form.[11] There is also another optical isomer, the D-form, which simply differs in how these groups are organised around the central carbon atom. The L-form which is built into our proteins slowly changes over, racemises, to the D-form until an equilibrium is reached where you have equal amounts of D- and L-forms. This is a reaction that depends on many things. It depends on temperature, on salt conditions, on pH, and so on. And it is different for every one of the nineteen optically active amino acids.[12] However, by pure chance one of these amino acids, aspartic acid, has a rate of racemisation that is almost identical to that of base-loss in DNA, which is probably one of the main causes of strand breakage when water is present. So we began to investigate whether the extent of aspartic acid racemisation correlates with the preservation of DNA. And it seems to do that. If we now look at remains where we can retrieve ancient DNA that we believe to be genuine we find that, for example in the Ice Man from the Austrian Alps that's around 5000 years old, most of the aspartic acid is in the L-form and very little in the D-form. When we look at 20 000-year-old horse remains from Chile, for which we cannot amplify any DNA, we find that we have quite a lot of D-aspartic acid relative to L-aspartic acid. And if we do this more systematically, in samples where we can retrieve ancient DNA, which in most cases have been reproduced in different laboratories, the D:L ratio for aspartic acid is always below 0.12.[6] In no remains that have a higher D:L ratio have we ever been able, despite several attempts, to find any DNA that actually lives up to our criteria for authenticity.

So this of course made it interesting for us to look into these cases where super-old DNA, and by that I mean things that are millions of years old, have been claimed to have been retrieved. The first that comes to mind is, of course, dinosaur DNA. So we looked at samples from four different dinosaur remains. One is a *Tyrannosaurus rex* from which DNA has been claimed to exist though it has not been published in the scientific literature. One is the dinosaur bone from Utah, and then there are two dinosaurs from Antarctica that we used with the idea that, for part of the time, they had been preserved under cold conditions so that they might have a higher chance of yielding DNA. But, in all these cases, the D:L ratio of aspartic acid is so high so we would not expect to find any DNA present. I think, if we believe this, it's quite unlikely that dinosaur fossils will ever yield DNA sequences.

The second group of fossils where super-old DNA has been reported are plant remains from Idaho where, in a lake bed 17 million years ago, leaves were embedded in clay.[7] These are marvellous fossils. When you split the clay open, the leaves are actually still greenish in colour but quickly turn black when they come into contact with oxygen. From these spectacular fossils, two DNA sequences, 1200 and 800 bp long have been reported. If we examine these leaf fossils for the aspartic acid isomers, so little remains that we can't determine a ratio. But when we looked at alanine, which is an amino acid that racemises more slowly than aspartic acid, so that we would always expect the D:L ratio to be lower than for aspartic acid, we find that the value is so high that we wouldn't expect any DNA to survive. This is not really surprising in view of the fact that water is present in these sediments. We know quite well the rate of depurination of DNA when water is present and can make some theoretical predictions. If we start from 10^{12} molecules, about the number you might expect in a gram of modern bone, and say that the pH was seven, the temperature was 15° C, and, making some other assumptions which are actually rather conservative in terms of rate of depurination, we would expect the last 800 bp fragment to suffer depurination and a strand-break after something like 5000 years. For an 80 bp fragment, it would be something like 50 000 years. So I believe there is a limit of something like 50–100 000 years on how far back you can go back and expect any molecules to survive in the presence of water.

That brings me on to the last group of super-old specimens there from which DNA sequences have been reported. These are the inclusions in amber which go back over 100 million years. You can find insects, plants and sometimes even vertebrates in amber. And if we now look at the D:L ratio for aspartic acid in amber inclusions, to my big surprise I must say, they are very low. They are all within the limits where we would still think of being able to retrieve DNA. So I think if we now want dinosaur sequences, what we really need to do is to wait for one to be found trapped in amber. In reality, I think that DNA from these amber inclusions have in the meantime proved not to be reproducible. Several groups have attempted to amplify DNA sequences from the same species from which they were originally reported and have had no success.[8] So I think there are probably other types of damage that take over in these amber inclusions. The fact that there is no racemisation is because there is absolutely no water in the inclusions. But gases will diffuse through amber and you probably have oxidative damage going on, putting a limit on how long DNA can survive even there. So I do think there probably is a time barrier, which I would put at between 100 000 and 1 000 000 years, that we will not be able to break with ancient DNA.

NOTES

1. Cross-striations refer to the normal microscopic appearance of muscle fibres where the molecular packing of actin and myosin filaments, the main muscle proteins, lead to a banding pattern that is picked up by histological dyes.
2. DNA is a double helix. The two strands are made up of a linear string of nucleotides or 'bases'. Each base pairs up with another on the opposite strand, so DNA lengths are usually given in base pairs, or bp for short. A 100 bp fragment is a double helix with a hundred bases on each strand.
3. This is the enzyme DNA polymerase which is used to amplify DNA from very small amounts by the Polymerase Chain Reaction (PCR).
4. As found in the cell nucleus. Mitochondria are found outside the nucleus and there are a thousand or more copies of mitochondrial DNA in a cell, making it easier to amplify than nuclear DNA where there are usually only two copies. Some nuclear genes, however, like ribosomal RNA genes, are present in multiple copies within the nucleus.
5. Where one base is changed to another.
6. Restriction enzymes cut the DNA helix at specific sites determined by the sequence. If this sequence is changed by a point mutation,

the enzyme will no longer be able to cut the DNA. This will show up as a change in the sizes of fragments seen after electrophoresis. There are hundreds of different enzymes available so there is a good chance that one can be found to detect any point mutation.

7. A much misunderstood term. It refers to the artificial insertion of single DNA molecules into bacteria. When the bacteria multiply, so does the introduced DNA. Taking the DNA from single bacterial colonies from a culture dish gives you enough DNA to sequence, all of it originating in the single molecule that got into the founder bacterium from which the colony arose.

8. The reference sequence is the first human mitochondrial DNA sequence to be published, in 1981.

9. The southern super-continent whose break-up led to the formation of Africa, America, Australia and Antarctica.

10. A large island between Australia and New Zealand.

11. The two forms, denoted D and L, are distinguished by the way solutions rotate polarised light—either to the left (laevo- or 'L') or to the right (dextro- or 'D'). The differences are due to molecular asymmetry.

12. Although there are twenty different amino acids in protein, one of them, glycine, does not have D- and L-forms because, lacking a side chain, there is no asymmetry around the central carbon atom.

7

Language and genes in the Americas

Ryk Ward

This paper will focus on a specific geographic region, the New World, and its original inhabitants, the Amerindians—a sub-group of our species that, in many respects, is quite recent in evolutionary terms. Despite this, Amerindians have developed a degree of cultural diversity which rivals that found in other, more ancient, regions of the world. They have also developed an appreciable degree of biological diversity. Thus, they lend themselves to an investigation of the extent to which the distribution of genetic diversity reflects the distribution of cultural diversity. More specifically, is it plausible to suppose that there is any relationship between the rate of cultural evolution and the rate of biological evolution?

It is first of all necessary to set the stage by examining the origins and overall antiquity of New World peoples. By looking at the archaeological evidence, as well as genetic and molecular evidence, we are able to draw together a number of disparate threads and identify some fairly consistent ideas about the timing and extent of colonisation. This will also shed light on how rapidly the different populations of the New World spread out and developed their own sets of cultures—including language. The latter half of the paper deals with the relationship between genetic diversity, largely as measured by molecular diversity in the mitochondrial genome, and the distribution of linguistic diversity in selected areas of the Americas.

7.1   AMERINDIAN ORIGINS

Irrespective of whether one believes colonisation took place relatively late (~14 000 years ago), or relatively early (~35 000 years ago), it is clear that humans entered the Americas well after populations, emanating from Africa, had colonised Asia, Europe and Australasia. The colonisation of the New World represents the last major thrust of the range expansion that anatomically modern humans have been engaged in for the past 100 000 to 15 000 years. Geographic isolation is the main reason that human occupation of the Americas lagged behind. The only land connection between the Americas and the rest of the world is the Beringian region. Even at the best of times, life in this area was very difficult and beset by extremes of cold and the other ecological constraints of high latitudes. Further, the first colonists had to conquer two important barriers before entry to the New World was assured.

The Bering Strait, a frigid stretch of inhospitable water between Siberia and Alaska, was the first—and most obvious—barrier. Less obvious, but no less important, was the barrier imposed by the periodic coalescence of the Laurentide and Cordilleran glaciers, two great ice sheets that dominated the ecology of North America during the Pleistocene (Fig. 7.1). Neither barrier was immutable but instead fluctuated in difficulty according to the waxing and waning of global temperatures during the closing phases of the Pleistocene. As the earth cooled, more water became locked up in ice and sea level dropped. Conversely, as temperatures warmed, ice sheets melted and the sea level rose. Consequently, the relationship between Asia and North America was dictated by the rise and fall of sea levels in response to the warming and cooling of the earth's climate. As temperatures cooled and the sea level dropped, the coastline extended with the result that in shallow areas such as the Bering Strait, there was continuity of land giving relatively easy access from eastern Siberia into the westernmost part of Alaska. However, the drop in temperature that caused the Beringian land bridge to emerge also resulted in the expansion and subsequent merging of the ice sheets, such that they formed an impenetrable barrier. Hence the first people to colonise the New

World had to deal with two opposing forces. The temperature had to drop sufficiently to expose the land bridge, so they could reach Alaska, but at the glacial maximum, when most land was exposed, entry to the rest of North America became impossible. Hence, the colonisation of the Americas should be viewed in terms of two distinct processes: first, the establishment of human habitation in western Alaska and second, the penetration of the ice sheets to reach temperate North America, after which little impediment existed to colonising the rest of both continents.

Since the rise and fall of sea levels have been dated quite reliably for the last 150 000 years or so, we can identify two periods of time when the Bering land bridge was open and humans could have

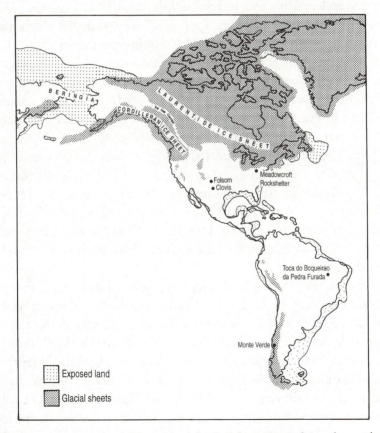

Fig. 7.1 The extent of the exposed land and continental ice-sheets about 18 000 years ago and the location of some early Amerindian sites.

reached Alaska. The first began around 50 000 years ago and lasted for approximately 15 000 years; the second, beginning some 25 000 years ago, lasted for approximately 10 000 years. In this context, it should be noted that a number of early archaeological sites have been identified in eastern Siberia that extend north, to around 60° latitude. In this high latitude region, there are some well-authenticated sites dating from 30–40 000 years ago, at a time when it was clearly cold and rather difficult to live there. During this period, there were certainly times when people could simply have walked across from Siberia into Alaska. So it appears that as far as anatomically modern humans were concerned, the initial colonisation of the tundra environment was not too problematic. They were certainly able to accomplish that 40 000 years ago. The issue is—how far did they get? Did they actually reach the Americas? From an archaeological standpoint, the evidence for reaching the Americas depends on identifying early sites in western Alaska, and although some putative early sites have been identified, none has really been fully substantiated. So we have a slight paradox in the sense that there are certainly early sites in eastern Siberia but, so far, no well-substantiated early sites in western Alaska. The earliest authenticated sites in western Alaska date from the period round about 12 000 years ago. At that time, the tundra was inhabited by large Pleistocene mammals and the people relied extensively on their ability to hunt big game both for food and other living necessities. This kind of environment had very little plant food that could be used to sustain even small human populations.

Whether the Bering land bridge was crossed during the first or second opportunity, by the time early humans first reached western Alaska, the two large ice sheets blocking the way to the rest of North America had most likely fused. It is worthwhile emphasising that these coalescing ice masses were not benign sheets that people could just walk across. Instead, early people would have had to deal with very rugged glacial ice that formed a virtually insurmountable barrier. The ability to traverse these large ice sheets would certainly not have been present for humans of 40 000 years ago or, for that matter, even 10 000 years ago. Hence, in the absence of a corridor between the glaciers, it would have been impossible for humans to reach temperate North America. Coastal voyaging is the only alternative. However, this option seems equally unlikely at times of maximum glacial extension since the coastline from southeastern Alaska to southern British Columbia was largely obliterated by formidable glaciers. Hence, a critical issue for dating the initial colonisation of the New World, and one that has largely been overlooked, is this: at what point were people able to travel through a narrow ice-free corridor and then populate the rest of North America? Here, I take issue with the concept that the ice-free corridor was a verdant valley through which people could easily walk. While we can only guess at the ecology, it was most likely a barren land, surrounded by tongues of glaciers and ice. So, even the ice-free corridor was a formidable challenge to early humans. This, in turn, implies a high degree of cultural evolution and sophisticated ability to survive under difficult conditions.

Entry around 15 000 years ago is entirely consistent with the large number of well-authenticated sites which date back to approximately 12 500 years. However, there are a smaller number of presumptive early sites, scattered around in South America, and one or two in North America, that have putative dates as old as 35 000 years. For example, the site of Monteverde situated in Chile was originally given a date of about 30 000 years. A second South American site, Pedra Furada in Brazil, has also been given a very early date—as has the very interesting site of Meadowcroft Rockshelter in Pennsylvania. However, if one actually goes and looks at a variety of early sites, the evidence starts to fade away a little bit. In Monteverde, for example, one finds that the stratigraphy is quite good back to about 12–13 000 years ago. Beyond that, the accuracy with which one is able to define human remains on the stratigraphic levels diminishes. In fact Tom Dillahay, who excavated the site, has in the last two or three years retracted his claim for its early age. He now thinks that deposits, which include pieces of wood which he believes are part of a shed or hut structure and are dated to approximately 13 000 years ago, in fact define the earliest levels. So the early site in Chile, at least as far as the person excavating it is concerned, appears to have become relatively recent.

Pedra Furada in Brazil is an open rock-shelter with a series of associated rock paintings. In fact it was the paintings that were the main reason for the initial excavation. Some of the pieces of painted rock have flaked off and have been found in fairly low levels. By fairly low, I mean levels which have been dated to 17 000 years. The issue is this: did the flakes fall off at that time or did they fall off later and then subsequently work their way down through the deposits? There are also a number of stone tools associated with Pedra Furada, but again, most people who have examined them believe that the evidence that justifies calling them manufactured tools, as opposed to stone that has splintered by falling from the cliff top, is not very strong. I think the consensus today is that Pedra Furada is not likely to be a very old site either.

On the other hand, sites which date back to some 12 000 years ago are both much more numerous and better characterised. The lithic industry characteristic of such sites is extensive and comprehensive and encompasses a number of unifacial and bifacial tools, besides the classic fluted points. There are also finely worked pieces of bone, presumed to have been used to craft spears and so forth. Not only are these Clovis[1] sites well characterised, but they are extremely numerous. There is also a great deal of continuity in the archaeological record for many of the sites. Further, there is the subsistence pattern practised by members of these cultures. Many sites are associated with large mammal kills, with stone artefacts, such as the fluted points so characteristic of Clovis and Folsom[2] cultures, often found in association with animals. There are very many instances where the remains of big game are found with spear points intact which attest to the success of the early humans as hunters. There are also butchery sites where literally hundreds of bison skeletons are found in association with projectile points, scrapers and other kinds of stone tools. There can be little doubt that by 12–13 000 years ago there was a relatively large human population successfully exploiting the large mammals that occurred throughout most of North America.

However, the most critical point is not that the archaeological record for the few putative early sites is suspect, but that there are essentially no sites at all for the period between 25 000 and 15 000 years ago. One characteristic of all human colonisation is that, once people get established, they tend to spread quite rapidly. Population numbers go up and the number of archaeological sites increases. If people really became established in the Americas as early as 30 000, or even 25 000 years ago, then there should be an increasing density of sites covering the next 10–15 000 years. This has never been observed, which is quite remarkable given the large number of American archaeologists who have looked for such evidence. To make the point even clearer, very soon after the most likely date of colonisation, some 15 000 years ago, there is an easily discernible proliferation of Clovis and Folsom sites through-out North America. This expansion appears to have occurred over a relatively short time period, approximately a thousand years. Similar expansions are also seen over the same time span in eco-logically equivalent areas of South America. So, ironically, the strongest arguments against the validity of the few sites reputed to be 30 000 years old is the absence of large number of well docu-mented sites from 18 000 years ago.

7.2   MOLECULAR CLUES TO AMERINDIAN ORIGINS

What about other kinds of data that might also shed light on Amerindian origins? It has been recognised for some time that genetic data can provide information about the overall evolution-ary relationships between population groups. In this respect, analyses of blood groups and other 'classical' genetic markers (serum proteins and red cell enzymes) have given considerable support to the notion that all Amerindians share a common source, and that this ancestral source had its roots in Asia. How-ever, these analyses rely on applying the principle of genetic drift to date population divergences and, under most situations, genetic drift only offers the ability to estimate population divergences over a time span corresponding to a few tens of generations. After that, the signals of more ancient historical affinities are essentially obliterated. Hence, gene frequency comparisons give relatively little detailed resolution about population affinities that are older than about 5000 years—or 200 generations. Fortunately, detailed comparison of DNA sequence data provides a way out of the impasse. This is because comparison of DNA sequences is essentially a means of counting up the number of mutations, or substitutions, that have accumulated since the divergence from a common ancestor. Provided we can make some reasonable assumptions about the rate at which mutations accumulate over time, the information about sequence differences can be translated into divergence times. This in turn yields an estimate of the time depth that separates populations. While this principle has been known for some time, it could not be put into practice until two obstacles were overcome. Firstly, more efficient molecular techniques had to be developed in order to generate extensive sequence information from large scale population samples and, secondly, a suitable piece of the human genome had to be identified for sequencing.

Just over ten year ago, the critical technical breakthrough occurred. This was the application of the newly invented Polymerase Chain Reaction (PCR) to generate the large amounts of DNA needed to act as the template for sequencing reactions. By using an *in vitro* strategy that relies on a recurrent series of DNA replication steps to produce a geometric increase in the amount of DNA from a targeted template, the PCR process is able to produce the quantities of DNA needed for accurate sequence reactions in an efficient and inexpensive manner. Once freed from the burdensome demands of using bacterial cloning to generate template DNA, sequencing efficiency increased exponentially. Like the PCR, sequencing strategies also rely on using DNA polymerases to replicate DNA in vitro, the essential difference being that in sequencing only a single round of replication is required but the reaction is carried out in four parallel steps—one for each of the four DNA bases. Each of these parallel reactions differs in that one, but only one, of the four bases is randomly replaced by a nucleotide analogue that inhibits further extension of the replicating DNA. Thus, each of the four reactions results in a mixture of fragments, all terminated by the appropriate base analogue and differing in size according to the position of the target base along the linear sequence of DNA. Separating the fragments by size and comparing their relative positions gives a representation of the full linear array of the nucleotides that make up the DNA strand being sequenced. The basic reason for needing the DNA amplification step is to generate a sufficient number of template molecules that the quantity of each of the resulting fragments is adequate to be visualised by the appropriate detection system—initially radio-activity, nowadays bioluminescent dyes.

However, even with these critical developments, the length of sequence that could be routinely generated for the relatively large samples needed for population studies, say 40–50 people per population, was still limited to 300–400 nucleotides. With such a relatively short sequence segment available for comparison and with the overall low rates of mutation, it became important to select a molecule that accumulated mutations fast enough to give an informative signal over the time spans relevant for human history. A decade ago, it was realised that the circular piece of DNA that forms the genome of mitochondria—the organelles that act as the power houses of cellular respiration—had a number of important advantages in this regard. The first is that the overall rate at which nucleotide substitutions accumulate in mitochondrial DNA appears to be roughly an order of magnitude greater than the rate at which they accumulate in nuclear genes. Moreover, the distribution of substitutions is not evenly spaced throughout the mitochondrial DNA molecule. The bulk of the approximately 16 500 nucleotides that make up the mitochondrial DNA molecule code for specific genes. However, the non-coding control region, of approximately 1000 nucleotides, is involved with the initiation of transcription and replication and it accumulates substitutions approximately ten times faster than the rest of the molecule. Even within the control region substitution rates are not uniform, and the first 350 nucleotides or so accumulate substitutions at an even higher rate. Sequence comparisons within this region, called hypervariable segment 1, or HVS 1, are the most informative over the short time spans typical of recent human evolution.

The second advantage of mitochondrial DNA is that it is inherited exclusively through the maternal line, which means its evolution has all the characteristics of haploid organisms. Specifically, the complicating features of genetic recombination can be ruled out, with the consequences that differences between sequences can be attributed to the overall accumulation of mutations during the time that the sequences have diverged from a common ancestor. As we see later, this means it is relatively easy to define the relationship between groups of sequences in term of a tree of ancestral relationships. The third advantage of mitochondrial DNA, which is often neglected, is a largely technical one. Unlike the situation for nuclear genes, where an individual cell has only two copies of each individual gene (one inherited from the mother, the other from the father), many copies of mitochondrial DNA can occur within a single cell. Cells may have a thousand or more mitochondria and each individual organelle has about ten circular DNA molecules. Hence for every single amplification target offered by a nuclear gene, a thousand or more targets are available from mitochondrial DNA.

The fundamental principle of Darwinian evolution is that not all individuals will have equal numbers of descendants, from which it follows that not all copies of a particular genomic segment will necessarily be transmitted to successive generations. An important corollary of this principle is that, for all copies of a particular genomic segment within a population, their ancestry will coalesce backwards in time to a single ancestral segment. The haploid transmission of mitochondrial DNA through the maternal line illustrates the point. In any reproducing population some women will have several daughters, some one, and some no daughters at all. The mitochondrial DNA of women who have no daughters is lost forever, while the mitochondrial DNA of women with many daughters is replicated many times. Thus, even when the population size remains constant, after a sufficient number of generations has passed, only a single mitochondrial lineage will have descendants and all the other lineages in the ancestral population will have been lost. Further, the random accumulation of nucleotide substitutions within the mitochondrial genome occur within the context of the coalescing lineages of ancestry. These independently occurring mutations can be used to construct a molecular phylogeny which can be considered an imperfect reconstruction of the underlying coalescent tree. Imperfect, because mutations are inherently unlikely to occur on more than a fraction of the branches of the coalescent tree, as opposed to occurring on all of them. Consequently, mutation rates that are slow relative to the true rate of ancestral coalescence will tend to give underestimates of coalescent time, and hence will underestimate the time to the most recent common ancestor. In this respect, even the rapidly mutating control region will tend to underestimate divergence times, since not all coalescent events will be identified. Sites within the control region with very high mutation rates tend to obscure the true number of mutations that have actually occurred because of recurrent changes at the same site. They too will exert a similar effect by underestimating divergence times, though how severe a problem this is has not yet been properly studied.

Since the coalescence of ancestry for a particular genomic segment sampled from a number of populations is contained within the coalescence of ancestry for the populations as a whole, an estimate of ancestral coalescence for a genomic segment is a good first approximation to the ancestral relationships between these populations. For the discussions which follow, it is important to note that molecular phylogenies are really the product of two independent processes. One has to do with ancestral coalescence itself, which is related to the demography of the population, and one has to do with the rate at which mutations accumulate. Since the loss of lineages is random through time, the rate at which ancestry coalesces backwards in time is a function of the demography of the population. Very small populations will have little genetic diversity and short, relatively shallow ancestral coalescent paths, while large populations exhibiting extensive molecular diversity will have much longer coalescent paths extending far back in time. The molecular phylogenies obtained from such populations will reflect these differences. Demographic changes also influence the distribution of ancestral coalescence, since populations that have undergone a rapid population expansion will have foreshortened coalescences compared to populations of constant size, and this will tend to result in a star-shaped molecular phylogeny. These points will become important later on when we evaluate the distribution of molecular phylogenies for contrasting groups of Amerindian tribes.

So what kind of data should we be looking at? We essentially want to construct a molecular phylogeny and from that molecular phylogeny we want to infer the population history. To do that, our first work was carried out in the Pacific North-west in the province of British Columbia. There are a number of reasons for choosing British Columbia for this kind of work. One of them has to do with the very high degree of linguistic diversity in this part of the world. Linguistic diversity in the Pacific North-west, in the sense of the number of languages spoken per square mile is considerably higher than virtually anywhere else in the Americas. Also, as you will see a little later, there are major linguistic differences amongst some of the tribal people that live in this area. The third reason is that many of the tribal groups that live in this area have had relatively little contact with the outside world. So it is easier to find individuals here who are truly representative of the tribes that existed 100 or 200 years ago than is the case for many tribes in the United States. For a number of different reasons, we chose to study the Nuu-Chah-Nulth, a tribe living on the west coast of Vancouver Island.

The Nuu-Chah-Nulth were certainly seen by Captain Cook over 200 years ago and their appearance recorded by one of the ship's artists. They also had some contact with the Spanish. However, the contacts with Europeans seem to have been fairly ephemeral and there was relatively little disruption of the villages and bands that make up the contemporary Nuu-Chah-Nulth. We used the village of Ahousaht as a base and decided on a sampling scheme that would take individuals from each of the fourteen major Nuu-Chah-Nulth bands and evaluate the distribution of sequence diversity.[1] Our initial sample was from 63 individuals. We ended up with a series of different lineages, where a lineage is defined as a sequence which differs from others at one or several different nucleotides. A particular lineage may occur either once or several times in the population. We can use the method of parsimony or a similar kind of statistical procedure to construct a molecular phylogeny. We can then look at the characteristics of that phylogeny and make some inferences about the kinds of evolution that have been required to generate this pattern of sequence diversity. The phylogeny that we constructed from the Nuu-Chah-Nulth data (Fig. 7.2) is characterised rather strikingly by the occurrence of distinct clades of lineages shown by the oval shading. Each clade is separated from the others by three to four mutational steps. However, there are some lineages which don't fit very well into any of the clades.

It is important to note that the characteristic feature of most Amerindian populations, which is really quite different from many other populations throughout the world, is the occurrence of these very well defined clades. And as you will appreciate from the data that Bryan Sykes presented, for example, one does not see such well-demarcated clades in European populations (Fig. 5.7, p. 111). So one of the issues is: how do the clades arise? Secondly, what is the distribution of these clades in a geographic sense? Our current interpretation is that each of the four clades represents the

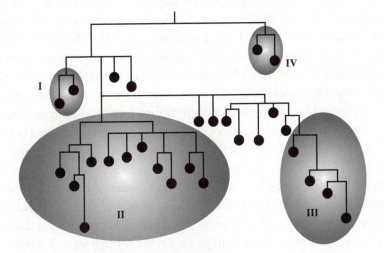

Fig. 7.2 A phylogenetic tree for 28 mitochondrial lineages of the Nuu-Chah-Nulth. The four lineage groups are enclosed by shading.

consequences of population isolation that occurred 40–50 000 years ago. This is because, using the presumed rate of sequence evolution, the divergence *between* clades tends to be approximately of that time depth. However, the molecular divergence *within* a clade is of the order of 10–15 000 years. So we have a situation where there were already different groups of molecular lineages in existence before the first colonists came to the New World. This may have come about by genetic differentiation due to the isolation of small populations in which mutations had accumulated, without very much migration between them. This was followed by the coming together, the admixture if you will, of descendants of each of those four different populations, and then subsequent evolution within the Americas. Now that may seem a fanciful story, but as you will see, there is actually quite a lot of data to back it up.

The first question might be: if this is what we observe in the Nuu-Chah-Nulth, what happens if we look at other tribes in North, and for that matter, Central and South America? Working with colleagues in Panama we compared the Nuu-Chah-Nulth and Haida—another Pacific North-west tribe—with the Ngobe from western Panama and, using published data, the Mapuche from Chile, and came up with the pattern of lineages seen in Fig. 7.3. This is not a tree but a network giving essentially the same information. The point I want to make from this diagram is that one sees a merging together or interdigitation of sequences from both South, Central, and North America, all in the same clades. So there are representatives or sequences of all four clades found throughout the Americas. You'll notice that the Ngobe from Central America lack sequences belonging to two of the clades. That seems to be generally true of other Central American populations in the sense that the degree of mitochondrial diversity is somewhat diminished compared to most other Amerind populations. Even more extensive surveys, using a slightly different system which picks up sequence variation all round the mitochondrial genome rather than just the control region show that, if we exclude Athabaskan and Aleut groups, essentially all Amerindian populations have lineages from all four clades.[2]

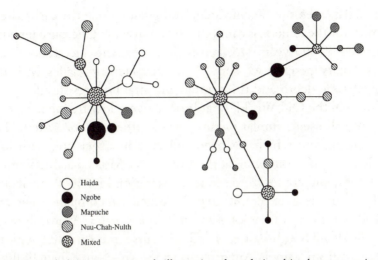

Fig. 7.3 A phylogenetic network illustrating the relationships between mitochondrial haplotypes in four tribal groups. The circles represent the haplotypes with areas proportional to their frequencies. (Data after Kolman et al.[5])

In any New World network, like that in Fig. 7.3, the lineages, or haplotypes as they are often called, at the centre of each clade, are also found in Siberia and even Mongolia. That comes back to the point I was making with respect to the Nuu-Chah-Nulth by reinforcing the supposition that the different clusters originally differentiated in the Old World and then migrated to the New World. So unless one wants to propose that the evolution occurred within the New World with back migration to Siberia and Mongolia, I think it becomes fairly clear that the distribution of diversity we see in the Americas is derived from two different kinds of processes. One is a process where the original lineages found in eastern Asia came in with the early colonists and still survive today. The other is the subsequent differentiation of lineages by mutations that have accumulated since the time of migration.

To come back to dates. The evidence is consistent in suggesting that the differentiation between clades dates back to differentiation due to genetic isolation that may be as old as perhaps 40–50 000 years, whereas the differentiation within Amerindian clades has a time depth of approximately 10–15 000 years. So the mitochondrial data seems to be consistent in suggesting that the early date of entry for the Americas cannot really be sustained, because if people arrived 30 000 years ago, and if they were the ancestors of contemporary people, we should see characteristic clades with that older time depth and the central, ancestral lineages should not be found in the Old World, or at least at a very low frequency.

We also see a similar picture with another genetic system. This is information which has been collected in association with John Clegg and his colleagues at the Institute of Molecular Medicine in Oxford, and comes from sequence variation in a 3000-nucleotide chunk of the beta-globin[3] gene.[3] When we compare a network made from a survey of 175 individuals from Africa, Europe, Central and South-east Asia and, of course, the Americas with the pattern from a single Amerindian tribe—the Nuu-Chah-Nulth— the essential point is that we see, in the Nuu-Chah-Nulth, representatives of all the major clusters that were defined in the global survey. Of course, the size and the degree of diversity within the Nuu-Chah-Nulth is considerably less, but it still suggests that the migrating population that formed their original ancestors had a fair degree of genetic diversity which is reflected by the retention of the various clades (see Fig. 7.4). That again is consistent with the concept that the four mitochondrial clades came in essentially as one migratory event. The beta-globin story is also rather interesting in that one can actually trace the origins of both the Nuu-Chah-Nulth and South-east Asian networks to a more fundamental network found in Mongolia. As I have already indicated, there are also a lot of mitochondrial lineages that are ancestral for Amerindians to be found in Mongolia as well. So Mongolia seems to be a key place of origin for the genes that ultimately found their way to the Americas. One further piece of evidence comes from a series of retroviruses that have become integrated into the human genome. When one compares the sequences of these HTLV-1 viral insertions in five British Colombians we find they are scattered on the tree drawn from the world-wide distribution. This is a very similar to the situation that we saw for the beta-globin. So my view is that the combination of mitochondrial data with the little bit of data that we have for the beta-globin, and the even

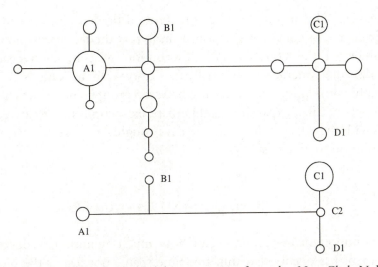

Fig. 7.4 Networks for beta-globin sequences from the Nuu-Chah-Nulth compared with a sample representative of global human diversity. Top network represents global pattern and bottom network is the Nuu-Chah-Nulth pattern: symbols identify evolutionarily distinct lineages. (Data after Harding et al.[4])

smaller amount of data that we have for HTLV-1 sequences, are consistently suggesting that the origin of Amerindian populations is relatively recent. By relatively recent I mean approximately 15 000 years ago or so, but that this original population was by no means a genetically depauperate. This was a population in which there was already a fair amount of diversity. However, perhaps we should think of not just one single large population walking across en masse, but a series of populations that may in fact have had no contact with each other, walking across independently over a period of 1–2000 years.

So as far as the entry into the New World is concerned, this is my take on the situation. First of all, I don't believe that there is any consistent information, archaeological, genetic or morphological, to suggest that people have been in the Americas for more than, say, 800 generations. As we'll come to in just a moment, people who speak Inuit and Na-Dene languages do tend to form a genetically distinct and fairly homogeneous cluster, and they are therefore different from the third major group, the Amerind speakers. The ancestral groups that crossed the land bridge had a fair degree of genetic dispersion. I think that the factors that have influenced the contemporary distribution of genetic diversity within the Americas is essentially due to two things. One is geographic proximity, the admixture between groups, and the other is the sets of tribal expansions and perhaps extinctions that followed. So with these factors in mind, we have roughly 14–15 000 years of evolution to play with.

7.3   GENETICS AND LANGUAGE

It is important to recognise just how much evolutionary development has occurred within this time span—not just at the biological but also at the cultural level. Early on there was already a well-defined set of different lithic industries. One can recognise in North America discrete cultural areas characterised by different linguistic affiliations and subsistence patterns ranging from hunting to the beginning of a reliance on cultivated plants and the onset of agriculture, with a corresponding influence on the overall distribution of population density. And while we don't have perhaps as much information as we would like, it is important to realise that large parts of North and South America had areas of very low population density prior to the arrival of the Europeans. However, some areas, particularly Central America and the lower part of North America, had areas of fairly high population density. So too did the Pacific coast of California and up into the Pacific Northwest itself, reflecting the richness of the ecosystem.

One of the most characteristic features of cultural development is the evolution of language and it is to that I wish to turn. Within the Americas there are a number of different major language groups. There is Na-Dene, which is spoken only in the northernmost parts of North America. There is Amerind, which is the language group for most Amerindian people, ranging from North through Central to South America. Then there is a third group which is spoken by Eskimos and Aleuts: the Eskimo-Aleut group. There are also two other language phyla that I want to draw your attention to because we will be discussing them right at the end. One is the Chukchi-Kamchatkan language, which is spoken today by the Chukchi who are essentially at the western extension of the Eskimos that today live in eastern Siberia. The second is the Uralic languages which are spoken over a wide swathe of northern Asia, and in particular by a group called the Yukaghir who are also immediately to the west of the Eskimo group. The point of making reference to those languages is to point out that Uralic and Amerind are thought to be, perhaps, part of a linguistic super-family, whereas Eskimo-Aleut, Chukchi-Kamchatka and Na-dene are very, very separate.

Once it was recognised that there were very different kinds of languages in the Americas, it was supposed that the occurrence of those different languages could be accounted for by postulating three distinct waves of migration, with each wave corresponding to a given language group. It was thought that the earliest migration might have occurred perhaps 12 000 years ago and would have given rise to Amerind speakers. A second migration might have occurred slightly later, giving rise to the Na-dene speakers, and then a third migration gave rise to the Eskimo–Aleut populations. One of the things that I will be addressing is whether or not the degree of genetic distinctiveness between these different linguistic groups, coupled with the presumption that this distinctiveness is a reflection of the very different times of entry, is mirrored in the data that we collect. In order to look at the distribution of linguistic diversity, it seemed to us that the most favourable area was, once again, in the Pacific North-west because here you have tribes who are geographically adjacent speaking either Amerind or Na-Dene languages. Our intention, then, was to see whether or not tribes in this part of the world, which belonged to these two very major different linguistic phyla, had an extensive degree of genetic differentiation.

The first study looked very simply at the smallest contrast we were able to make, and still have something meaningful to say. We chose the Wakashan-speaking Nuu-Chah-Nulth who, you will recall, live on the west coast of Vancouver Island, and the Salishan-speaking Bella Coola from the mainland. Both are Amerind languages and Salishan is almost a sister-language of Wakashan, the difference between the two being more or less equivalent to that between French and Italian. The third point of contrast was the Haida, Na-Dene speakers who live on the Queen Charlotte Island to the North.[4] If the hypothesis of linguistic divergence and differential migration were correct, we should expect to see a major separation between the Haida on the one hand and the Bella Coola and Nuu-Chah-Nulth on the other. We would expect very different kinds of mitochondrial lineages in the Haida compared to those in the Amerind speakers. The reality was quite different. Figure 7.5 shows a molecular tree of all the lineages found in the three tribes, with a symbol indicating in which tribe each lineage was found. Wherever you look you see that there is considerable interdigitation between lineages. In other words, we do not find any correspondence between lineage and linguistic affiliation. That tends to cast considerable doubt on the potential relationship, or the potential separation, of the two language groups.

Obviously, three populations is too small a sample for general conclusions. What I want to show you now are the results of a

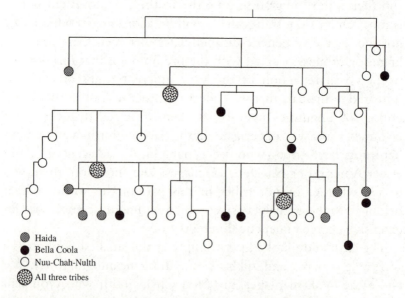

Fig. 7.5 Molecular phylogeny, based on a maximum likelihood tree, of samples from three tribes from the Pacific North-west.

much more extensive study in which we have included, in addition to the Haida, Bella Coola and Nuu-Chah-Nulth, a third Amerind-speaking tribe, the Yakima—from the state of Washington. We add to this a Na-Dene Athapaskan group and two Eskimo–Aleut populations: the Inuit from Greenland and Yu'pik from eastern Siberia and western Alaska. We are also able to include representatives of three other language families from North-east Asia: the Chukchi, the Altai and the Uralic-speaking Yukaghir. We now have five different linguistic phyla in our study (Fig. 7.6). Although the time scale here is arbitrary, it is possible, I suppose, that the Amerind and the Uralic, perhaps the Altaic groups, might have some tenuous connection, perhaps 12–15 000 years ago. By contrast, the Na-Dene, Chukchi–Kamchatka, and Eskimo–Aleut languages are acknowledged to be quite distinct. From the genetic point of view, we might expect to have a fairly close relationship between Yu'pik and Inuit speakers, a rather distant relationship between Haida and interior Athapaskan speakers, and an intermediate kind of relationship between the three Amerind tribes.

However, when we look at the data, what we see is totally different (Fig. 7.7). The Altai do in fact form an out-group, with the Uralic-speaking Yukaghir also being a distinct group. So the

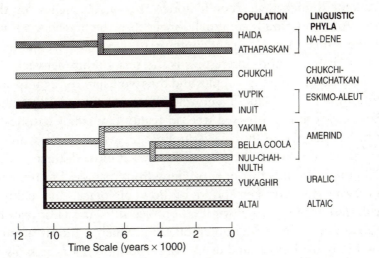

Fig. 7.6 A linguistic phylogeny of circumarctic populations.

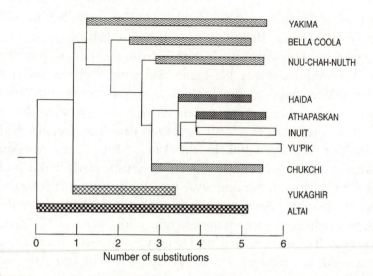

Fig. 7.7 A molecular phylogeny of circumarctic populations using the same shading as for Fig. 7.6.

Altai and the Yukaghir do more or less what we might expect. However, everything else then changes. As far as the mitochondrial DNA phylogeny is concerned, we have a clustering of lineages which essentially group together the populations of four different linguistic phyla: Amerind, Na-Dene, Eskimo–Aleut and Chukchi–Kamchatkan. You'll notice that the branches for the three Amerind groups are much longer than the branches for any of the other groups, again contrary to what we would have expected. Even more interesting is the relationship between the two Na-Dene speaking samples and the two Eskimo–Aleut speaking samples. They cluster very closely together, and although I have drawn this as a rooted tree, in reality that is not statistically justified, and we really have four lineages which form a multifurcation with the Chukchi being somewhat more distinct.

So the inferences that one can draw from that are, I think, very interesting for our perception of how biological and cultural evolution occurs. At the first level of inferences, it is clear that the relative rate of biological differentiation, which is mostly a function of population size and degree of isolation, is not necessarily at all close to the rate at which languages can evolve. In other words, if we assume that population isolation and differentiation will give rise to the inexorable accumulation of different mutations at, on average, a fairly steady rate, the same is not necessarily true for languages. This particular diagram suggests that Na–Dene and Eskimo–Aleut, two very different languages, could perhaps have sprung into being very rapidly over quite a short period of time. What's relevant here is that these populations tend to be subsisting on a very marginal ecological resource base. They are very small populations that tend to be widely separated and perhaps that's where you might imagine linguistic differentiation would occur most frequently.

Having said all that, we do have to remind ourselves that we have only just begun. The vast bulk of the data that I've shown you, for the reasons I explained near the beginning, is all attributable to information that derives from mitochondrial DNA. In order to define population history properly, we really need to have a series of pieces of information—a series of molecular phylogenies—from many, many different independent genes. I will leave you with the thought that the information thus far suggests that the relationship between biological differentiation and linguistic differentiation is incredibly complex—and therefore extremely interesting. But I also want to leave you with the realisation that we've really only scratched the surface, and that there is probably twenty times as much work still to be done.

NOTES

1. Named after the site of Clovis in New Mexico.
2. Named after the site of Folsom in North Dakota.
3. One of the subunits of haemoglobin.

8

Human genetic diversity and disease susceptibility

Walter Bodmer

8.1 INTRODUCTION

The emphasis in my paper is on the sources of genetic variation in human populations, and its implications. Where does the genetic variation come from? Why is it there? What maintains it in populations? I aim to relate these questions to the variation in populations which has been the subject of previous chapters, and also to include some discussion on mutation which is the ultimate origin of variation between genes in any population.

Identical twins are real human clones, unlike Dolly the sheep,[1] because, having a common mother, they have the same mitochondria. One of the striking things about identical twins is how very similar they look at all ages from a few months, to a few years, to the age of 70 and beyond. This extraordinary similarity at different ages is due to the fact that identical twins share the same identical genetic contribution from both their parents. Their similarities tell us that outward features by which we recognise people really must have a very large genetic component. Apart from identical twins, we all differ from one another in our facial features so there must be an enormous amount of variation, or polymorphism, for those features. There are special regions of the brain for facial recognition and one of my hopes is that, in my final years as a scientist in Oxford, I might be able to work out the genetics of the face, which is to me a most fascinating subject. Then one could reconstruct the face from DNA. From a small sample of

DNA you could perhaps produce a picture of a supposed criminal for the newspapers the following day.

In contrast to the similarities shared by identical twins, there are enormous differences between almost all other people. My wife and I saw this most vividly in the contrast between ourselves and the pygmies of Central Africa, whom we went to study with Luca Cavalli-Sforza in the late 1960s. This was one of the earliest expeditions for the study of human polymorphisms using the new techniques for preserving the white cells of the blood. The main aim was to study the relationship of the pygmies to other African populations. Their extraordinary differences from the average European, in skin colour, height, hair texture and many of those features by which we usually categorise people in different population groups, are the outward features which are most likely to have been subject to the evolutionary pressures of natural selection. A light skin is a marked disadvantage in a tropical climate and reflecting this is the fact that the incidence of skin cancer in light-skinned people in the tropics is several hundred-fold higher than that for dark-skinned people. Light skin presumably evolved in the higher latitudes of the northern hemisphere under conditions of relative lack of sunlight, in order to provide enough vitamin D.[2] Whether this happened before the beginning of the last Ice Age, or during it while human populations were forced to move south, or during the short period of time between the end of the Ice Age and the spread of agriculture from the Middle East is not yet clear. As to the difference in overall size, large, ungainly people would not fare well in the tropical rain forests of Central Africa and there is, in fact, a general correlation for comparable species between reduced size and a tropical forest habitat.

There is no doubt about the huge amount of variation between individuals, both within and between populations. At least some of this variation, including the outward features by which we tend to recognise each other, has probably been subject to quite strong natural selection. When we study the fine chemical differences between individuals, such as are revealed by the blood groups, or such as can now be established from the overwhelming extent of differences at the DNA level, it can be shown that by far the majority of the genetic variation between people is within any population rather than between populations. It is the *frequencies* of these genetic variations in different populations which can be used to characterise them.

8.2   MUTATION AS THE SOURCE OF GENETIC VARIATION

The ultimate source of all genetic variation is mutation, namely the generation of a novel DNA sequence, either through changing a single base pair or many bases. Usually this is the result of mistakes either in the process of DNA replication, the effects of chemicals on DNA, or the effects of external radiation. The rate at which mutations occur is very low, perhaps of the order of 10^{-9} or 1 in 1000 million base pairs per generation. That is an estimate of the rate at which the average single base pair of DNA changes but the rate may be very different and much higher for particular types of sequences. However, on average, the overall mutation rate within any given gene is likely to be somewhere between 1 in a million and 1 in 100 000. Most mutations, when they occur, are deleterious if they have a functional effect. If they are sufficiently deleterious and completely knock out the function of the gene, they may lead to a severe disadvantage especially when they are homozygous. Thus, there are many rare inherited diseases in populations which are maintained by a balance between mutation and selection, classically analysed by Danforth and Haldane. The simple relationship between the frequency, p, of a gene maintained at a low level by the balance between a mutation rate m and selective disadvantage s can be represented by the formula:

$$2p = m/s \qquad (1)$$

Consider, for example, the inherited susceptibility to colon cancer, or *polyposis coli*, which occurs in the population with a frequency of between 1 in 5000 and 1 in 10 000. It can be estimated that its selective disadvantage is approximately 40%, which means that under 'primitive' conditions, corresponding to those of premodern times, the ability of such individuals to reproduce would only be about 60% of the norm. Using formula (1), the mutation rate would lie between 1 in 12 500 and 1 in 25 000 which is very much on the high side. If the selective disadvantage was only 20%, then these rates would be halved, namely, a mutation rate between 1 in 25 000 and 1 in 50 000. Such rare inherited variants, which are maintained by the balance between mutation and selection, do not, individually, make a major contribution to genetic variation in populations as a whole, but collectively they contribute to an overall frequency of clear-cut severe inherited disease of as much as 1%.

At the other end of the spectrum of natural selection, we think of the normal evolutionary process as the production of occasional advantageous mutations which have an increased fitness, either in terms of fertility or viability, or perhaps through mating preferences. From time to time such mutations will occur in any given gene, and will then sweep through a population because of their selective advantage. Although this advantage might be quite small, it nevertheless results in genetic change in the population which can be rapid when compared to an evolutionary time-scale. Subsequently, there may be another mutation in the same gene which again gives a selective advantage, so that the normal evolutionary process for any particular gene is one of successive substitutions of one advantageous mutant by another. However, at any given time the probability of polymorphism, namely of more than one mutation being present at an appreciable frequency in the population, is quite small, because the process of spread is rapid relative to the time interval between their appearance.

Mutations which are neutral, namely, which have neither a selective advantage nor disadvantage, may increase in frequency just by chance. This process was first analysed by R.A. Fisher and Sewall Wright and, more recently, strongly promoted by the late Moto Kimura. It was originally thought that neutral mutations would be rare, but now that we know there are many regions of the DNA where mutations, if they arise, are not likely to be associated with any selective advantage or disadvantage, this is no longer the case. Thus, mutations occurring in sequences between

genes or in the intervening sequence between exons,[3] and at positions, often the third base-pair position,[4] which do not give rise to
a change in the amino acid sequence in a coding region are all
likely to be neutral in their effects. Although the chance of any
given neutral mutation increasing in frequency is small, there are
so many opportunities for them to arise that many of the differences now found within populations may indeed be neutral.
Models for the changes in frequency associated with neutral mutations show that it is possible for there to be several of them present
in a population at any one time. The increase in the frequency of
neutral mutations is determined by their mutation rate and the size
and structure of the population and not by the type of gene in
which the mutation occurs.

8.3 BALANCED POLYMORPHISMS

A classic case, almost unique in human populations, for which the
basis for natural selection is understood is the disease sickle cell
anaemia and its partial manifestation, the sickle cell trait. This
anaemia was first described by an American physician in Chicago,
James Herrick, in 1904. He had the insight to realise that the
anaemia presented by a West Indian patient from Grenada was out
of the ordinary. From this observation, a long and convoluted
story eventually led Linus Pauling and his colleagues to establish
sickle cell anaemia as the first disease whose molecular basis was
fully understood. The sickle-shaped red blood cells, which are the
hallmark of the anaemia, are due to an abnormal form of haemoglobin.[5] Sickle cell anaemia arises when an individual has two
genes for the abnormal form of the haemoglobin and can make no
normal haemoglobin. Carriers of the abnormal haemoglobin gene
have just one version of it, their other version coding for the
normal haemoglobin, and they are clinically quite normal. However, following standard Mendelian[6] genetics, when two carriers
mate, on average a quarter of their offspring will inherit two
copies of the abnormal gene and so suffer from sickle cell disease.
Another quarter will have two normal copies and make only

normal haemoglobin, while half the offspring will again be carriers with one normal and one abnormal copy. Under the prevailing environmental conditions in West Africa, where the sickle cell gene seems to have originated, individuals with sickle cell anaemia usually die when they are very young. This explains the puzzling observation that the frequency of the anaemia in African black populations is much lower than might be expected. They simply do not survive. The sickled red blood cells can also be seen in the carriers under conditions when oxygen is limiting and their frequency in some parts of West Africa can be as high as 20%. How then can there be such an extraordinarily high frequency of an apparently deleterious gene, where matings between carriers are almost bound to lose nearly a quarter of their offspring to sickle cell anaemia?

There are now many other similar examples of inherited abnormalities connected with red blood cells, and in particular with its major component, haemoglobin, which is responsible for carrying oxygen around the body. One class of these diseases is thalassaemia in which the haemoglobin, rather than being abnormal, is made in insufficient quantities. Thalassaemia is found particularly in the Mediterranean basin, which accounts for its name.[7] The famous geneticist, J.B.S. Haldane, who lived on the site of Wolfson College in Oxford as a child, noticed a clear relationship between the distribution of thalassaemia and that of malaria. He suggested, in the early 1950s, that the reason why thalassaemia was so common may be that carriers of an abnormal gene, who themselves were clinically normal, might have a selective advantage with respect to resistance to malaria. This idea was obviously also applicable to carriers of the sickle cell gene. Tony Allison, who also worked in Oxford, was the first to show clearly that the distribution of the sickle cell gene throughout Africa could be almost exactly superimposed on the distribution map of endemic malaria. There is now an accumulation of evidence to support the notion that the reason for the high frequency of the sickle cell gene in parts of Africa, and other genes such as that for thalassaemia in other parts of the world, is because of a balance between the advantage to the carriers with respect to resistance to malaria, and the disadvantage of the homozygotes who suffer from a severe, often lethal anaemia.

The simple algebra which explains how selective advantage and disadvantage can lead to a balanced polymorphism, in which two versions of a gene are maintained stably in a population, was first worked out by another of the great pioneers of evolutionary genetics and my first, great teacher, R.A. Fisher, in 1922. He was the originator of much of our modern understanding of the quantitative basis of evolution and also a founding father of modern statistics. He coined the term 'variance' in one of his earliest papers and the idea of randomisation, which is fundamental to properly designed clinical trials, was also his. The basis for Fisher's model and his calculation is shown in Table 8.1. It can be assumed that the relative fitness of the sickle cell gene carrier, AS, is 1, and that this fitness is greater than that of the normal individual, AA, because of the advantage with respect to malaria by an amount s. The disadvantage of the homozygote for the abnormal haemoglobin gene, SS, is measured by t. When the anaemia is lethal, meaning that no individuals survive to reproduce, the value of t is 1, giving a fitness of 0. Fisher derived the simple result that under these conditions at equilibrium, the frequency of the S gene is $t / s + t$ while that for A is $s / s + t$. If the anaemia is lethal and t is 1, a frequency of 20% for the carriers could be explained by a 25% advantage for the sickle cell heterozygote, AS, over the normal homozygote, AA, presumably due to the resistance to malaria. If, as another example, the relative fitness of the individuals with sickle cell anaemia were 50%, then the selective disadvantage for the normal homozygotes would be 12.5% instead of 25%. The population frequency of carriers can then be used, on the basis of this model, to estimate the selective advantage of the sickle cell

Table 8.1 Equilibrium gene frequencies under heterozygote advantage

	AA	AS	SS
If genotypes			
have fitnesses	$1 - s$	1	$1 - t$
then, at equilibrium, alleles	A	S	
will have frequencies	$t/(s + t)$	$s/(s + t)$	

Note: For heterozygote advantage, both s and t must be positive heterozygote with respect to malaria given an estimate of the selective disadvantage for those with sickle cell anaemia. This must have been close to 1 under the conditions in West Africa prevalent at the time the sickle cell gene polymorphism evolved. Nature, therefore, plays a cruel bargain with humanity in maintaining this balance between the resistance to malaria of certain genetic combinations, and the severe, almost lethal, disease of others.

In the Mediterranean area where thalassaemia is relatively common, there are many different types of the disease caused by different mutations in the same gene. These occur with different frequencies in different populations and so can be used to some extent to characterise the differences and similarities between populations. A few years ago when I was involved in making a 'Horizon' programme for BBC television about population studies, I had the opportunity to interview a Greek Orthodox Cypriot priest in London, Father Andreas, and his son Father Constantine. They knew all about thalassaemia because Father Andreas had had three sons with the disease, two of whom had died. I pointed out to them that the frequency of thalassaemia variants in Greek and Turkish Cypriots was essentially identical, and markedly different from the comparable frequency distributions amongst people on the mainland of Greece or the mainland of Turkey. This clearly indicates that the Greek and Turkish Cypriots are part of the same population—a Cypriot population—which is neither Greek nor Turkish. The Greekness of the Greek Cypriot population must therefore have been acquired culturally. Father Andreas and his son were truly astonished to hear this because they had believed unequivocally that their Greek culture meant that they were indeed Greeks. I explained to them that that was most probably not the case. Their Greek culture has most probably been acquired by a process called 'elite dominance' which refers to the situation where a small number of people, usually aggressive invaders, impose their culture but do not contribute many of their genes to the population that they have so strongly influenced with their culture. Fathers Andreas and Constantine responded by saying that 'Perhaps if only people understood that, it might help deal with the political problems.' If only that were true! It seems, on the contrary, that wherever there are political problems between population groups, they are often between those peoples that are most similar to each other genetically.

Another example of a genetically determined disease which has an interesting distribution throughout Europe is Rhesus disease of the newborn. The genetically determined Rhesus blood groups fall into two simple categories, namely positive and negative. Rhesus negative individuals lack a molecule on the surface of their red blood cells which is present in all those who are Rhesus positive. When a woman who is Rhesus negative has Rhesus positive children, the difference is recognised by the immune system. By the time of the second, third and further pregnancies, if they are Rhesus positive, the mother's immune response causes severe damage to the foetal red cells, attacking them almost as if they were an infection and giving rise to a severe disease called Rhesus haemolytic disease of the newborn.[8] Fortunately, this disease can now be treated prophylactically[9] so that it is now no longer a major problem in this country. A striking feature of the distribution of Rhesus negativity is that it is highly focused in Northern Europe, with frequencies of up to 16%, especially in the Basque country. Its frequency in Southern Europe is quite low so that Rhesus disease was really a genetic disease limited to the populations of Northern Europe.

It was the distribution of Rhesus negativity across Europe that led Luca Cavalli-Sforza to make his initial suggestion about the consequences for gene frequency distributions across Europe of the migration of farmers outwards from the Middle East across Europe, starting some 10 000 years ago. The high frequency of Rhesus negativity in the Basques is also of considerable interest, since there is evidence to suggest that the Basque people may, to a considerable extent, reflect features of the pre-Neolithic populations found in Europe before the advance of agriculture from the Middle East. What can explain the high frequency of Rhesus negativity in some populations when it is undoubtedly associated with a disease that has negative selective consequences? Some years ago at Stanford, with colleagues Marc Feldman and Markus Nabholz, we looked at models of this process to see what sort of selective balance might compensate for the disadvantage of the haemolytic disease of the newborn and so give rise to the observed polymorphism. The theoretical study simply says that the distribution can be explained if there is some compensatory advantage for being Rhesus negative or carrying one gene for Rhesus negativity, but why that should be so is not at all clear. It could also be the case that the increase in Rhesus negativity in the Basque and other Northern European populations was simply a chance phenomenon associated with the very low population densities that must have existed in Europe before the advance of agriculture. In a population which is predominantly Rhesus positive, the Rhesus negative gene will not be at a disadvantage due to haemolytic disease of the newborn until it has increased in frequency sufficiently for there to be an appreciable frequency of Rhesus negative individuals, who need to carry two copies of the Rh^- gene. By contrast, in a population which was predominantly Rhesus negative, a Rhesus positive individual, who need only have one copy of the Rh^+ gene, would be at an immediate disadvantage.

Colour blindness is a common normal genetic variation. It occurs in about 7% of males in Europe, but only about half a percent of females. This is because the genes which code for the colour vision pigments, and which are defective in colour blind males, are found on the X chromosome. Males, who are XY, have a defective gene on their one X chromosome but have no second, normal chromosome to compensate for this. On the other hand, females, who are XX, can have a defective gene on one of their X chromosomes, such as for colour blindness, but so long as their other X chromosome is normal, they will not express the defect but just be carriers. The simple Hardy–Weinberg Law[10] of population genetics applied to the X chromosome tells us that if the frequency of colour blindness in males is 7%, its frequency in females will be the square of this, namely 0.49%. Colour blindness has no obvious disadvantage, though it may affect visual acuity which could have been important during the time when human populations were hunters and gatherers. It is a puzzle, however, why it is relatively common, for example, in Western Europe but comparatively rare in most African populations. When my wife and I went with Luca Cavalli-Sforza to study the pygmies in the Central African tropical rain forests and administered the Ishihari test for colour blindness, they outlined the shape of the outside of the circle, which was a sensible response to our request, though not the one anticipated. However, it was clear that the frequency of colour blindness, if it was present at all amongst the pygmies, was very low. Why then should it be so high in other populations? Has there been some selection for other aspects of visual acuity in colour blind people, for example, following the development of agriculture? There are data to suggest that there exist variants of the pigment genes which do not give rise to defective colour vision, but influence colour perception. Could that account for differences in colour perception between males and females, since males with their single X chromosome are more likely to express such variation than are females with their two X chromosomes? Though colour blindness has no disease connotations whatever, it may nevertheless influence opportunities for work. Though you may have a pilot's licence if you are colour blind, you cannot have a commercial airline pilot's licence and in Japan, I understand, it may even be difficult to get an ordinary driver's licence. That seems unreasonable to us, but nevertheless the example of colour blindness emphasizes the fact that it may sometimes be relevant to know about genetic variation in relation to different types of employment.

8.4   THE HLA SYSTEM AND AUTOIMMUNE DISEASE

The autoimmune diseases arise when the body's immune system, which normally recognises the differences between our own tissue and foreign invaders, malfunctions and attacks our own tissues. The role of the immune system is to protect us from infections by viruses, bacteria, moulds such as yeast, and parasites such as that which causes malaria. The chronic autoimmune diseases, which arise when the immune system malfunctions and attacks self, are relatively common and include, for example, rheumatoid arthritis and juvenile onset insulin-dependent diabetes. These troublesome diseases have a major genetic component, but are not inherited according to simple Mendelian principles in the same way as sickle cell disease or *polyposis coli*. The understanding of the genetic basis for these sorts of diseases came from a direction that might seem surprising, namely from the attempts to find the genetic differences that are the cause of tissue transplant rejection.

It is well known that you cannot take a small piece of skin from anyone else and put it as a graft onto your own skin and have it survive for any appreciable length of time. The foreign skin will be rejected because of the widespread genetic differences between individuals. These are recognised by the body as foreign and lead to rejection by the immune system, just as an infection is rejected. This is the reason why it is not possible to perform organ or bone marrow transplants from one individual to another without doing something to prevent rejection of the donor organ by the recipient's immune system. There are two fundamental approaches to this problem, both of which are used in clinical practice. The first is to find drugs which inhibit the function of the immune system, and the second is to search for the genetic differences that are recognised when a tissue graft is rejected in the hope that matching for these differences will improve graft survival.

The search for the genetic systems underlying graft rejection followed closely the principles which led to the discovery of the red cell blood groups such as ABO and Rhesus. It has led, in work that I and my wife Julia Bodmer have been involved in for more than 30 years, and through major international collaborations, to the description of an extremely complex system of genetically determined differences that are present on the white cells of the blood. These white cells, such as lymphocytes, carry out most of the functions of the immune system.

The major genetic system that was discovered is called HLA— 'H' for human, 'L' for leucocyte[11] or lymphocyte, and 'A' for what was thought to be the first of such systems. HLA, in fact, defines a small genetic region on human chromosome 6 which contains many different genes, of which seven are relevant for graft rejection because they are the genes that vary most between individuals. The genes fall into two classes with somewhat different functions. The first (class I) are called A, B and C and the second (class II) are DRB, DQA, DQB, DPA and DPB. The variants of the A gene are simply identified by the letter A followed by a number, and similarly for each of the others. The number of variants so far identified at each of the different loci in the HLA region, including many that are much less variable, are listed in Table 8.2. Each of the HLA protein products is actually formed from two protein chains. For the class I products, one chain comes from one of the A, B or C genes and the second from another gene, β2-microglobulin, which does not vary and is coded for on a different chromosome. The class II products are formed from two chains, both coded for within the HLA genetic region. Thus, the DR products are formed from the DRB chain and a DRA chain which does not vary, while the DQ and DP products are formed from combinations of the DQA and DQB, or DPA and DPB chains. In these latter two cases, both of the chains forming the DQ and DP products can be quite variable. There are hundreds of variants at these different HLA gene loci[12] generating a huge amount of genetic variability between individuals. Many people have spent a long time defining and investigating their distributions in different populations and in different diseases.

Because the HLA genes lie close together on the chromosome they tend to be inherited *en bloc* (Fig. 8.1). Each of us carries two blocks of HLA genes, one version on the chromosome 6 inherited from our father and the other from our mother. The versions of the genes we inherit together on a chromosome from one parent are called collectively a 'haplotype'. On average, following Mendel's Laws, each haplotype is passed on, mostly intact, to one half of the offspring. However about 2% of the haplotypes passed on to our offspring will be formed from recombinations between the two parental HLA haplotypes each of us has inherited. Most of the time, however, a mating will produce just four different possible combinations of HLA types, so that there is a chance of only one quarter of finding a pair of siblings who are exactly matched for their HLA types. The fact that grafts exchanged between siblings matched for HLA survived far longer than those between unmatched siblings provided some of the first and most

Table 8.2 Number of alleles described at different loci within the HLA region

Locus	Alleles
HLA-A	60
HLA-B	125
HLA-C	36
HLA-E	4
HLA-F	1
HLA-G	5
HLA-H	1
HLA-I	1
HLA-J	1
HLA-K	1
HLA-L	1
HLA-DRA	2
HLA-DRB1	132
HLA-DRB2	1
HLA-DRB3	5
HLA-DRB4	4
HLA-DRB5	6
HLA-DRB6	1
HLA-DRB7	1
HLA-DRB8	1
HLA-DRB9	1
HLA-DQA1	16
HLA-DQB1	25
HLA-DQA2	1
HLA-DQB2	1
HLA-DQB3	1
HLA-DOB	1
HLA-DMA	4
HLA-DMB	4
HLA-DNA	1
HLA-DPA1	8
HLA-DPB1	63
HLA-DPA2	1
HLA-DPB2	1

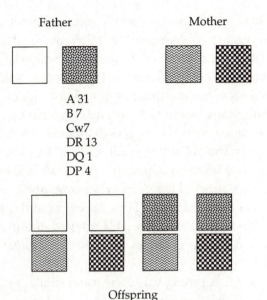

Father   Mother

A 31
B 7
Cw7
DR 13
DQ 1
DP 4

Offspring

Fig. 8.1 Inheritance of HLA types in blocks. The combination A31, B7, Cw7, DR13, DQ1, DP4 is passed on *en bloc* most of the time to the children. Each square, two different ones for the father and two different ones for the mother, represents such a combination of six types. Each parent passes on one or other of the two combinations with a probability of one half. Thus, there are four possible combinations of offspring, each of which occur with the same frequency of nearly one quarter. The frequency with which combinations are broken due to genetic recombination is about 2%.

striking evidence that these HLA types were critical for transplant survival. Nevertheless, a good level of survival, even between HLA matched sibs, will not be achieved without the use of drugs which suppress the body's immune responses. Only grafts exchanged between identical twins will survive in the absence of such drugs, showing that there are still important genetic differences, other than HLA, which matter for graft survival. The improvement in the effectiveness of immuno-suppressive drugs has meant that the difference in graft survival between matched and unmatched individuals, at least in the case of kidney transplantation, has been markedly reduced, but there is still an important contribution from the effects of matching.

The situation with respect to bone marrow grafts is somewhat different. There it has proved much more difficult to achieve effective survival in the absence of an HLA match. Bone marrow grafts are used to treat some rare inherited enzyme-defect diseases and also certain types of leukaemia and lymphoma. Taking a bone marrow sample is not as traumatic as donating a kidney and so it is quite straightforward, when there is an HLA-matched sibling, to use his or her bone marrow for transplantation. However, especially because of the relatively small sizes of families nowadays, most individuals who need a bone marrow graft will not have an HLA identical sibling. That is why organisations such as the Anthony Nolan Bone Marrow Trust have accumulated panels of up to half a million people who are HLA typed in order to try and find as good a match as possible, amongst a volunteer panel, for those people.

The original HLA typing was done using similar techniques to those that led to the discovery of the red cell blood groups, and before the relevant genes were identified at the DNA level. Nowadays, all HLA typing is done using DNA-based technology. This depends on the remarkable so-called polymerase chain reaction or PCR technique.[13] Using the polymerase chain reaction (PCR), small defined segments of any source of DNA can be amplified in the test tube and easily evaluated for their sequence. As a result, all the HLA types can now be easily identified using a minute sample of DNA such as can be obtained from a small amount of blood or from scraping cells from the mouth, or perhaps the few cells attached to the root of a pulled hair.

Studies with the mouse equivalent of the HLA system, called H2, had suggested that they might be involved in certain types of disease susceptibility. The question was soon asked: are there any associations between any of the human HLA types and auto-immune diseases, such as rheumatoid arthritis? These studies were accelerated when it came to be realised that the HLA types were critically involved in key aspects of the control of the body's immune responses. The genetic basis for an autoimmune disease might involve some difference in the way the body responds inappropriately to its own tissue, or perhaps is stimulated to do so following an infection.

Ankylosing spondylitis is an autoimmune disease, sometimes called 'poker spine', where the immune system appears to attack the tissue between the vertebrae. This leads to fusion between the vertebrae and a most unpleasant stiffening of the spine, which is, however, not generally life-threatening. The striking observation was made nearly 25 years ago that virtually all people with ankylosing spondylitis have the marker HLA-B27, whose frequency in a normal European population is well under 10%. It is, however, important to emphasise that only a few per cent of people with B27 get ankylosing spondylitis. On the other hand, if you do not have B27 then you almost certainly will not get the disease. That is what is meant by a genetically based susceptibility. It is an important contribution to the risk of getting the disease which can be helpful in its diagnosis or treatment, but it is not the only factor. In the case of ankylosing spondylitis in women, where the disease is less severe, Julia Bodmer, in a study with the Hills many years ago,[1] showed that it was helpful to do B27 typing because this led to an earlier diagnosis of ankylosing spondylitis in women who could then be helped with appropriate physiotherapy. Other diseases that have been clearly associated with HLA types include rheumatoid arthritis, insulin-dependent juvenile onset diabetes and coeliac disease— a sensitivity to gluten, a component of wheat and some other cereals. Intriguingly, among the diseases associated with HLA is haemochromatosis, which is not thought to be an autoimmune disease. Recently, however, haemochromatosis has been shown to be associated with a mutation in a gene neighbouring the HLA region, and that probably explains its association with the HLA marker A3. That leads to the question: what other genes are there around in the HLA region, and how does their close linkage affect the evolution of genetic variants within a population?

8.5   SELECTION, LINKAGE, AND LINKAGE DISEQUILIBRIUM IN THE HLA REGION

The HLA region, if defined as the segment of chromosome 6 which encompasses all the HLA class I and II genes, has been shown to be about 4 million base pairs long, and so constitutes just over 1% of the total human genome. However, the HLA genes themselves take up only a part of this space on chromosome 6. Between the class I and class II gene clusters are many other genes, only a few of which may have anything to do with the immune system. In addition, genes have been found outside this region, on its HLA-A side, which bear some evolutionary relationship to the HLA-A, B and C genes, but are clearly different from them. It is here that a variant has been found which seems to account for the genetic determination of the disease haemochromatosis.

The question to be asked is if, for example, A3 is indeed strongly associated with haemochromatosis, is it actually that variation in the A gene which is the cause of the genetically based disease susceptibility or could the effect be due to a difference in a closely linked gene? In that case we now know, because the haemochromatosis gene has been identified, that the A3 association is simply acting as a marker to point out the genetic region where the true explanation for the disease susceptibility lies. To understand the general answer to that question we need to probe further the pattern of HLA distributions in populations and their interpretation. Of course, much of the information on the distribution of HLA variations in human populations has been collected because of the medical interest in the system, both with respect to transplantation matching and because of its association with disease.

As shown in Table 8.2, there are many different genes of the HLA region that can be included alongside HLA-A, B and C and HLA DRB etc. These are genes whose sequence similarity is sufficiently close to one or other of the two classes of HLA genes that they can be clearly defined as belonging to them. Thus, HLA-F and G are functioning genes very similar to HLA-A, B and C. Why then should A and B have so many variants, C have significantly fewer and yet G have hardly any at all? There are also enormous differences in the number of variants found for the other related genes in human populations. If all the variation was simply due to mutation, with no selective effect connected with the functions of the genes, there would be no reason why any one of them should have more variants in the population than any other. There is no evidence for assuming that the mutation rate differs to any significant extent between the different genes. The only possible explanation, therefore, for such differential variability within and between genes is that natural selection is at work.

Another related line of evidence for natural selection comes from looking at the distribution of the position of the variants within the HLA genes in relation to the known structure and function of the molecules, which was first worked out by Bjorkman and Wiley and their colleagues some ten years ago.[2] We now know that the function of these molecules is to hold peptides, short fragments of proteins containing from 8 to 15 amino acids, in a clearly formed cleft in order to present them to the immune system. Most of the polymorphic variants in the HLA genes occur at amino acid positions in and around the cleft. From this it is presumed that they have something to do with the efficiency with which the protein fragments can be trapped in the cleft and therefore the differential efficiency with which they can be presented to the immune system. The pattern of variation within the gene is also highly localised. Polymorphisms are found surrounding the cleft—exactly where one would expect to find them if they are to be functionally relevant. That could only arise through the action of natural selection. Thus, it is most likely that the extensive HLA polymorphism exists because the different variants have different effects on immune responses which are important in protecting us from infection. The variety of types may be due to the fact that human populations have been subjected to a wide variety of different infections at different times, each generating their own particular new selective needs.

Detailed examination of the patterns of variation between alleles shows that they are often clustered, and that new patterns of variation can be derived from old by recombination. Alleles, which are versions of the complete gene, can thus be thought of as combinations of these variable clusters or *epitopes*. New alleles are derived from old by recombining these epitopes to give, presumably, differential effectiveness in resistance to different infections. The system of variants has been selected for during evolution to withstand the onslaught of a variety of different infections. The autoimmune diseases, which today are associated with particular HLA variants, are the price we pay now, as we live longer and suffer from this variety of chronic diseases for having withstood all these different infections in the past.

The classic explanation, given earlier, for the maintenance of a balanced polymorphism such as for the sickle cell gene, was Fisher's simple model for the selective advantage of the heterozygote over both homozygotes (see Table 8.1). Such an explanation does not easily fit the extensive observed polymorphism of the HLA system. However, it can be shown that a fluctuating situation, in which the heterozygote is sometimes at an advantage over one homozygote but not the other, and vice versa, can also give rise to a balanced polymorphism. This is exactly as might be expected from the need to withstand different infections at different times. Sometimes one type is at an advantage and sometimes another type. Maintaining variability in the population preserves the variations that may be needed to withstand different infections as they come along. Thus, I believe that the HLA polymorphisms have been subject to very strong evolutionary pressures, which are enormously variable in time and throughout human populations.

8.6   LINKAGE DISEQUILIBRIUM—THE POPULATION ASSOCIATION BETWEEN CLOSELY LINKED ALLELES

The HLA-A and B genes do not behave independently in populations, because they are closely linked on chromosome 6. For example, in a typical British population, the proportion of people who are *A1* is 31%, and of those who are *B8*, 21%, but the proportion who are *A1* and *B8* is 18%. This is nearly three times that expected if *A1* and *B8* were independently distributed in the population, namely 0.3×0.21 or about 6.5%. That excess frequency of the combination *A1B8*, over what is expected assuming independence, is a measure of the association between these two types in the population. It is explained by the fact that the genes on the

same chromosomes, namely the haplotypes, also show an association between *A1* and *B8*. Thus, while the separate frequencies of
the alleles *A1* and *B8* are, respectively, 0.17 and 0.11, so that the
product is expected to be less than 2%, the actual observed frequency of the chromosome combination or haplotype *A1B8*
found in the population is nearly 10%. That is an association at the
genetic level, which has been called linkage disequilibrium. This
association is absolutely key to understanding the behaviour of
closely linked genes in a population and now underlies most of the
approaches being used for working out genetic contributions to
multifactorial disease.

The simple theoretical basis for linkage disequilibrium between
closely linked alleles was worked out by Jennings in 1917 and is
illustrated below.

Haplotype: *AB*   *Ab*   *aB*   *ab*
Frequency: $P_A.P_B + D$ $P_A.P_b - D$ $P_a.P_B - D$ $P_a.P_b + D$
$D^{(n)} = (1-r)^n.D^0$
If r = 0.1% (roughly 10^5 base pairs) then, after 1500 generations
$D^{(1500)} = (1-0.001)^{1500}.D^0 = 1/5.D^0$

The four possible combinations of two alleles at two closely
linked loci would, in the absence of any population association,
occur with frequencies that are simply the products of frequencies
of the component alleles. The quantity D, which measures the
extent to which pairs of alleles tend to occur together, is also a
measure of the departure of the frequencies of the four haplotypes
from what is expected in the absence of association. For example,
if D is large, then there is an excess of types *AB* and *ab* over the
alternatives *Ab* and *aB*. Jennings devised the simple formula for the
rate at which the association between alleles, D, declines. He
showed that D is reduced by an amount $(1-r)$ each generation and
so by an amount $(1-r)^n$ after *n* generations. This means that D
tends towards 0 with increasing number of generations, and at an
increasing rate as the recombination fraction (*r*) increases. Thus,
the further two genes are apart, the more rapidly alleles at them
will lose their association. For example, with a recombination
fraction of 0.1%, representing on average 100 000 base pairs, the

linkage disequilibrium, D, would go down by a factor of 5 in 1500 generations which, allowing about 25 years per human generation, is getting on for 40 000 years. That, for example, is about the length of time that *Homo sapiens* has been in Europe. Thus, associations between alleles in populations can be related to the known patterns of human migrations.

The phenomenon of linkage disequilibrium makes haplotypes very useful markers for human population studies. The frequencies of haplotypes in different populations can give much more information about population relationships than the frequencies of the individual genes on their own. For example, the combination *A1B8DR3* on the same chromosome is relatively common in Northern Europe. It perhaps represents a type that was present in Mesolithic, pre-Neolithic populations that must, at some time, have populated these northern parts of Europe before the advance of agriculture. The haplotype *A1B8DR3* is also an extraordinarily good marker of European migration to other parts of the world, so it is inevitably found at a relatively high frequency for example, in Australia, Canada and the United States. Its frequency in the United States is slightly less because there were many migrants who went there from Central and Southern Europe. The combination *A1B8DR3* is not at all common in Southern Europe. *A1* alone is not only a marker for Europeans, but is found almost uniquely in all Caucasoid populations, including those in India. It is possible to find other combinations that are, for example, distinctive of African populations or Oriental populations, and the data clearly show that these haplotype distributions are the most distinctive HLA frequency markers for characterising different human populations.

8.7 MULTIFACTORIAL DISEASE INHERITANCE

Frances Galton, Charles Darwin's cousin, first promoted the idea, in the second half of the 19th century, that twins could be used for assessing the inherited components of quantitative differences between individuals. He pointed out, before Mendel's work was rediscovered, that the difference in concordance for a disease in identical and non-identical twins, was a measure of genetic contribution. For example, if you have one member of a twin pair identified as hypertensive, the frequency with which the other member of the twin pair is hypertensive is the concordance, and for identical twins that is about 40%. The concordance figure is much lower for non-identical twin pairs, showing the extent to which hypertension may be determined by genetic factors. Similar results can obtained for mental deficiency, manic depressive states and even tuberculosis, where the genetic contribution may be through immune response differences.

How can one look for the genetic components to such relatively common diseases? That in itself would need to be the subject of another paper. However, the principle of linkage disequilibrium leading to population associations with genetic markers is a powerful tool for finding genetic components to common diseases. Given a candidate gene that, due to its function or perhaps position on the chromosomes, appears to be relevant for hypertension, the approach is to find a common variant in or near that gene and then ask the question: does that variant differ in frequency in people with and without hypertension? If there is such a difference, then that is reasonable evidence that somewhere in the neighbourhood of the gene there exists a mutation which may confer susceptibility to hypertension much in the same way that B27 is a susceptibility gene for ankylosing spondylitis. However, for such studies it is very important to have good control populations that can provide the basis for comparison with marker distributions in diseased individuals.

8.8   THE INHERITANCE OF NORMAL TRAITS

The same approach can be used for studying any individual differences, including those which control light skin colour and complexion. There is a clear gradient in skin colour throughout Europe, from light in the north to somewhat darker in the south, which must have evolved as an adaptation to low sunlight. It is striking how little is known about the genetics of light skin colour, even though it is undoubtedly the major genetic determinant of skin cancer. The major opportunities for the evolution of light skin colour are, as already mentioned, either after the end of the last Ice Age, some 12 000 or so years ago before the agricultural migrants came to Northern Europe from the Middle East, or after the arrival of the migrant *Homo sapiens* some 40 000 years ago. There is, perhaps, at most a window of opportunity of about 20 000 years during which light skin colour and complexion could have evolved. Is the lighter skin complexion of Oriental populations due to mutations in similar genes or is it quite independent in its origins?

Variation in the genes controlling facial features might surely give the most marked differences between relatively closely related populations. Is there male choice of the female mate in the human populations and, if so, is the male somehow entrained to his mother's facial features, giving rise to a form of assortative mating? There are other examples of high levels of variation whose origin is not yet well understood. One example is that in olfactory receptors, the smell receptors which determine most of the subtlety of taste differences. There are very complex and variable genes controlling such differences and there is no doubt that we all differ in our taste preferences. That perhaps is the molecular basis for the well-known aphorism '*de gustibus non est disputandum*'.[14]

NOTES

1. In 1996, scientists in Edinburgh succeeded in producing Dolly the sheep, the first animal of any size to have been cloned from an adult body cell. In this case, the nucleus of a mammary cell grown in culture was transferred into a donor egg that had it own nucleus removed. The egg was implanted into a recipient 'surrogate' mother. Only the mitochondria, which are outside the nucleus, survive from the donor egg.
2. Vitamin D is converted to its active form by the action of ultraviolet light on the skin.

3. The genes of higher organisms are made up of exons, which contain the coding information, and intervening sequences, the introns, which do not code for proteins.
4. The genetic code is written in threes where one triplet of bases specifies a particular amino acid. Because of redundancy in the system, sets of similar triplets can code for the same amino acid and, in this case, it is the third base which usually varies. So mutations can happen at these bases without changing the amino acid being specified by the code.
5. Haemoglobin is the protein which carries oxygen in the blood. It is found in the red cells and gives blood its colour.
6. After Gregor Mendel who discovered the basic rules of genetics in the middle of the 19th century.
7. From thalassa (Gk): the sea.
8. Haemolytic meaning that the red blood cells are lysed.
9. By injecting the mother with antibodies just after the birth of the first Rhesus positive baby. The antibodies destroy any Rh$^+$ cells which have entered the mother's circulation thereby preventing an immune response from getting under way.
10. The basic law of population genetics, independently arrived at by the English mathematician G.H. Hardy and the German physician Weinberg in 1908, which explains why dominant and recessive genes are maintained at equilibrium frequencies in the population.
11. The generic name for white blood cells.
12. A genetic locus (pl. loci) refers to a position on the chromosome.
13. See Chapter 7 for a description of this technique.
14. In matters of taste there is no dispute.

Bibliography

CHAPTER 1

1. Cann, R. L., Stoneking, M., and Wilson, A. C. (1987). Mitochondrial DNA and human evolution. *Nature* **325**, 31–6.
2. Darwin, C. (1859). *On the origin of species.* John Murray, London.
3. Huxley, T. H. (1865). On the methods and results of ethnology. *Fortnightly Review* **1**, 257–77.
4. Wainscoat, J. S., Hill, A. V. S., Thein, S. L., Flint, J., Chapman, J. C., Clegg, J. B., and Higgs, D. R. (1989). Geographic distributions of alpha- and beta-globin gene cluster polymorphisms. In *The human revolution*, (ed. P. A. Mellars and C. B. Stringer). Edinburgh University Press, Edinburgh.
5. Jones, Sir W. (1786). Third anniversary discourse: 'On the Hindus'. Reprinted in *The collected works of Sir William Jones III*, (1807) pp. 23–46. John Stockdale, London.
6. Trubetzkoy, N. S. (1939). Gedanken über das Indogermanenproblem. *Acta Linguistica* **I**, 81–9. Reprinted in *Die Urheimat der Indogermanen* (ed. A. Scherer, 1968), pp. 214–23. Wissenschaftliche Buchgesellschaft, Darmstadt.
7. Greenberg, J. H. (1963). *The languages of Africa.* Indiana University Press, Bloomington.
8. Greenberg, J. H. (1987). *Languages in the Americas.* Stanford University Press, Stanford.
9. Ruhlen, M. (1987). *A guide to the world's languages, volume 1: classification.* Stanford University Press, Stanford.
10. Childe, V. G. (1926). *The Aryans.* Kegan Paul, Trench and Trubner, London.
11. Gimbutas, M. (1980). The Kurgan wave migration (c. 3400–3200 B.C.) into Europe and the following transformation of culture. *Journal of Near Eastern Studies* **8**, 273–315.
12. Renfrew, C. (1987). *Archaeology and language, the puzzle of Indo-European origins.* Jonathan Cape, London.
13. Ammerman, A. J. and Cavalli-Sforza, L. (1973). A population model for the diffusion of early farming in Europe. In *The explanation of culture change: models in prehistory* (ed. C. Renfrew), pp. 335–58. Duckworth, London.
14. Zvelebil, M. and Zvelebil, K. V. (1990). Agricultural transition, 'Indo-European origins', and the spread of farming. In *When worlds collide: the*

Bellagio papers (ed. T. L. Markey and J. A. C. Greppin), pp. 237–66. Karoma, Ann Arbor.

15. Ehret, C. (1988). Language change and material correlates of language and ethnic shift. *Antiquity* **62**, 564–73.

16. Bellwood, P. (1991). The Austronesian dispersal and the origins of languages. *Scientific American* **265**, 88–93.

17. Higham, C. F. W. (1994). Archaeology and linguistics in southeast Asia: implications of the Austric hypothesis. Paper presented at the 15th Indo-Pacific Prehistory Association Congress, Thailand.

18. Nichols, J. (1992). *Language diversity in space and time.* University of Chicago Press, Chicago.

19. Excoffier, L., Pellegrini, B., Sanchez-Mazas, A., Simon, C., and Langaney, L. (1987). Genetics and the history of sub-Saharan Africa. *Yearbook of Physical Anthropology* **30**, 151–94.

20. Cavalli-Sforza, L. L., Piazza, A., and Menozzi, P. (1994). *The history and geography of human genes.* Princeton University Press, Princeton, New Jersey.

21. Richards, M., Côrte-Real, H., Forster, P., Macaulay V., Wilkinson-Herbots, H., Demaine, A., Papiha, H., Hedges, R., Bandelt, H-J., and Sykes, B. (1996). Paleolithic and Neolithic Lineages in the European Mitochondrial Gene Pool. *American Journal of Human Genetics* **59**, 185–203.

22. Barbujani, G., Pilastro, A., De Domenico, S., and Renfrew, C. (1994). Genetic variation in North Africa and Eurasia: Neolithic Demic Diffusion vs. Paleolithic Colonisation. *American Journal of Physical Anthropology* **95**, 137–54.

23. Torroni, A. Stringer *et al.* (1992). Native American mitochondrial DNA analysis indicates that the Amerind and the Na-Dene populations were founded by two independent migrations. *Genetics* **130**, 153–62.

CHAPTER 2

1. Thorne, A. and Wolpoff, M. H. 1992. The multiregional evolution of humans. *Scientific American* **266**: 76–83.

2. Frayer, D., Wolpoff, M., Smith, F., Thorne, A. and Pope, G. 1993. The fossil evidence for modern human origins. *American Anthropologist* **95**: 14–50.

3. Wolpoff, M. and Caspari, R. 1997. *Race and human evolution: a fatal attraction.* Simon and Schuster: New York.

4. Stringer, C. 1992a Replacement, continuity and the origin of *Homo sapiens*. In (G. Bräuer and F. Smith eds) *Continuity or replacement: controversies in* Homo sapiens *evolution*, pp. 9–24. A. A. Balkema: Rotterdam.

5. Stringer, C. and Bräuer, G. 1994. Methods, misreading and bias. *American Anthropologist* **96**: 416–424.

6. Grün, R. and Stringer, C. 1991. Electron spin resonance dating and the evolution of modern humans. *Archaeometry* **33**: 153–199.

7. Grün, R., Brink, J., Spooner, N., Taylor, L., Stringer, C., Franciscus, R. and Murray, A. 1996. Direct dating of Florisbad hominid. *Nature* **382**: 500–501.

8. Braüer, G., Yokoyama, Y., Falgueres, C. and Mbua, E. 1997. Modern human origins backdated. *Nature* **386**: 337–338.

9. Braüer, G. 1992. Africa's place in the evolution of *Homo sapiens*. In (G. Braüer and F. Smith eds) *Continuity or replacement? Controversies in* Homo sapiens *evolution*, pp. 83–98. A. A. Balkema: Rotterdam.

10. Aiello, L. 1993. The fossil evidence for modern human origins in Africa: a revised view. *American Anthropologist* **95**: 73–96.

11. Stringer, C. and Gamble, C. 1993. *In search of the Neanderthals*. Thames and Hudson: London.

12. Hublin, J.-J., Spoor, F., Braun, M., Zonneveld, F. and Condemi, S. 1996. A late Neanderthal associated with Upper Palaeolithic artefacts. *Nature* **381**: 224–226.

13. Hublin, J.-J., Barroso Ruiz, C., Medina Lara, P., Fontugne, M. and Reyss, J.-L. 1995. The Mousterian site of Zafarraya (Andalucia, Spain): dating and implications on the palaeolithic peopling processes of Western Europe. *Compte Rendu Hebdomadaire des Séances de l'Académie des Sciences, Paris*. IIa, **321**: 931–937.

14. Wood, B. 1992. Origin and evolution of the genus *Homo*. *Nature* 355: 783–790.

15. Stringer, C. 1996. The Boxgrove tibia: Britain's oldest hominid and its place in the Middle Pleistocene record. In (C. Gamble and A. J. Lawson, eds) *The English Palaeolithic reviewed*, pp. 52–56. Trust for Wessex Archaeology.

16. Bermudez de Castro, J. M., Arsuaga, J. L., Carbonell, E., Rosas, A., Martinez, I. and Mosquera, M. 1997. A hominid from the Lower Pleistocene of Atapuerca, Spain: possible ancestor to Neandertals and Modern Humans. *Science* **276**: 1392–1395.

17. Arsuaga, J.-L., Martinez, I., Gracia, A. and Lorenzo, C. 1997. The Sima de los Huesos crania (Sierra de Atapuerca, Spain). A comparative study. *Journal of Human Evolution* **33**: 219–281.

18. Ruff, C. 1994. Morphological adaptation to climate in modern and fossil hominids. *Yearbook of Physical Anthropology* **37**: 65–107.

19. Holliday, T. 1997. Body proportions in Late Pleistocene Europe and modern human origins. *Journal of Human Evolution* **32**: 423–448.

20. Swisher, C., Curtis, G. and Jacob, T. 1994. Age of the earliest known hominids in Java, Indonesia. *Science* **263**: 1118–1121.

21. Swisher, C., Rink, W., Anton, S., Schwarcz, H., Curtis, G., Suprijo, A. and Widiasmoro 1996. Latest *Homo erectus* of Java: potential contemporaneity with *Homo sapiens* in Southeast Asia. *Science* **274**: 1870–1874.

22. Gibbons, A. 1996. *Homo erectus* in Java: a 250,000 year anachronism. *Science* **274**: 1841–1842.

23. Fullagar, R., Price, D. and Head, L. 1996. Early human occupation of northern Australia: archaeology and thermoluminescence dating of Jinmium rock-shelter, Northern Territory. *Antiquity* **70**: 740–773.

24. Brown, P. 1992. Recent human evolution in East Asia and Australasia. *Philosophical Transactions of the Royal Society, London, Series B* **337**: 235–242.

25. Pardoe, C. 1993. Competing paradigms and ancient human remains: the state of the discipline. *Archaeology in Oceania* **26**: 79–85.

26. Stringer, C. 1992b. Reconstructing recent human evolution. *Philosophical transactions of the Royal Society, London, Series B*, **337**: 217–224.

27. Lahr, M. 1996. The evolution of modern human diversity: a study of cranial variation. Cambridge University Press: Cambridge.

28. Stringer, C., Humphrey, L. and Compton, T. 1997. Cladistic analysis of dental traits in recent humans using a fossil outgroup. *Journal of Human Evolution* **32**: 389–402.

29. Nei, M. 1995. Genetic support for the out-of-Africa theory of human evolution. *Proceedings of the National Academy of Sciences USA* **92**: 6720–6722.

30. Gibbons, A. 1997. Ideas on human origins evolve at Anthropology gathering. *Science* **276**: 535–536.

31. Rogers, A. and Jorde, L. 1995. Genetic evidence on modern human origins. *Human Biology* **67**: 1–36.

32. Takahata, N. 1995. A genetic perspective on the origin and history of humans. *Annual Review of Ecological Systematics* **26**: 343–372.

33. Krings, M., Stone, A., Schmitz, R., Krainitzki, H., Stoneking, M. and Paabo, S. 1997. Neandertal DNA sequences and the origin of modern humans. *Cell* **90**: 19–30.

34. Ward, R. and Stringer, C. 1997. A molecular handle on the Neanderthals. *Nature* **388**: 225–226.

CHAPTER 3

Allen, W. Sidney. 1978. *Vox latina*. 2nd ed. Cambridge: Cambridge University Press.

Allen, W. Sidney. 1987. *Vox graeca*. 3rd ed. Cambridge: Cambridge University Press.

Bibliographie linguistique. Published by the Permanent International Committee of Linguists. (Last vol. 1994.) Dordrecht: Kluwer.

Brunner, Karl. 1965. *Altenglische Grammatik*. 3rd ed. Tübingen: Niemeyer.

Campbell, Alistair. 1959. *Old English grammar*. Oxford: Clarendon Press.

Goddard, Ives. 1974. 'An outline of the historical phonology of Arapaho and Atsina.' *International journal of American linguistics* **40**.102–16.

Godel, Robert. 1975. *An introduction to the study of Classical Armenian.* Wiesbaden: Reichert.

Greenberg, Joseph H. 1963. *The languages of Africa.* Bloomington: Indiana University Press.

Greenberg, Joseph H. 1987. *Language in the Americas.* Stanford: Stanford University Press.

Hübschmann, H. 1897. Armenische Grammatik. I. Theil: Armenische Etymologie. Leipzig: Breitkopf and Härtel.

Jespersen, Otto. 1948. *A Modern English grammar on historical principles.* Part I, 2nd ed. Copenhagen: Munksgaard.

Kökeritz, Helge. 1953. *Shakespeare's pronunciation.* New Haven: Yale University Press.

Kökeritz, Helge. 1961. *A guide to Chaucer's pronunciation.* New York: Holt, Rinehart and Winston.

Labov, William. 1994. Principles of linguistic change. Vol. 1: internal factors. Oxford: Blackwell.

Mossé, Fernand. 1952. *A handbook of Middle English.* Translated by James A. Walker. Baltimore: Johns Hopkins Press.

Nichols, Johanna. 1990. 'Linguistic diversity and the first settlement of the New World.' *Language* **66**.475–521.

Nichols, Johanna. 1996. 'The comparative method as heuristic.' M. Ross and M. Durie (eds.) **1996**:39–71.

Ringe, Don. 1992. 'On calculating the factor of chance in language comparison.' *Transactions of the American Philosophical Society,* Vol. **82**, Part 1.

Ringe, Don. 1995. 'The 'Mana' languages and the three-language problem.' *Oceanic linguistics* **34**.99–122.

Ringe, Don. 1996. 'The mathematics of 'Amerind'.' *Diachronica* **13**.135–54.

Ringe, Don. Forthcoming. 'Probabilistic evidence for Indo-Uralic.' To appear in a conference volume edited by Brian Joseph and Joe Salmons and published by Benjamins.

Ringe, Don. In progress. 'The problem of binary comparison.'

Ross, Malcolm, and Mark Durie (eds.). 1996. *The comparative method reviewed.* Oxford: Oxford University Press.

Schmitt, Rüdiger. 1981. *Grammatik des Klassisch-Armenischen mit sprachvergleichenden Erläuterungen.* Innsbruck: Innsbrucker Beiträge zur Sprachwissenschaft.

Sturtevant, Edgar H. 1940. *The pronunciation of Greek and Latin.* Philadelphia: Linguistic Society of America.

Wells, J. C. 1982. *Accents of English.* Cambridge: Cambridge University Press.

Wolfe, Patricia. 1972. *Linguistic change and the great vowel shift in English.* Berkeley: University of California Press.

CHAPTER 4

1. Kauffman, S. A. (1993) *The Origins of Order*. Oxford University Press.
2. Dover, G. A. (1982) Molecular drive: a cohesive mode of species evolution. *Nature* **299**: 111–117.
3. Jackson, M., Strachan, T. and Dover, G. A. (1996) *Human Genome Evolution*. Bios. Scientific Publishers, Oxford.
4. Dover, G. A. (1992) Observing development through evolutionary eyes: a practical approach to molecular coevolution. *Bioessays* **14**: 281–287.
5. Gray, S., Szymanstic, P. and Levine, M. (1994) Short-range repression permits multiple enhancers to function autonomously within a complex promoter. *Genes and Devel.* **8**: 1829–1838.
6. Arnone, M. I. and Davidson, E. H. (1997) The hardwiring of development: organization and function of genomic regulatory systems. *Development* **124**: 1851–1864.
7. Chothia, C. (1994) In Akam, M., Holland, P., Ingham P. and Wray, G. (eds) *The Evolution of Developmental Mechanisms*. *Development* Supplement, Cambridge.
8. Bork, P. (1992) Mobile Modules and Motifs. *Curr. Opin. Struct. Biol.* **2**: 413–421.
9. Miklos, G. L. G. and Rubin, G. M. (1996) The role of the genome project in determining gene function: insights from model organisms. *Cell* **86**: 521–529.
10. Raff, R. A. (1996) *The Shape of Life*. University of Chicago Press.
11. Bateson, W. (1894) *Materials for the Study of Variation*. Macmillan and Co., London.
12. Garciá-Bellido, A. (1975) Genetic control of wing disc development in *Drosophila*. *Ciba Found. Symp.* **19**: 161–182.
13. Holland, P. W. H. and Garcia-Fernandez, J. (1996) Hox genes and chordate evolution. *Dev. Biol.* **173**: 382–395.
14. Akam, M., Holland, P., Ingham P. and Wray, G. (eds) (1994) *The Evolution of Developmental Mechanisms*. *Development* Supplement, Cambridge.
15. Halder, G., Callaerts, P. and Gehring, W. J . (1995) Induction of ectopic eyes by targeted expression of the *eyeless* gene in *Drosophila*. *Science*. **267**: 1788–1792.
16. Garciá-Bellido, A. (1996) Symmetries throughout organic evolution. *Proc. Natl. Acad. Sci. USA* **93**: 14229–14232.
17. Scott, P. M. (1994) Intimations of a creature. *Cell* **79**: 1121–1124.

CHAPTER 5

1. Herzfeld L, Herzfeld H (1919) *Lancet* **ii**: 675–678.

2. Boyd W (1950) *Science* **112**: 187–197.
3. Quicke DLJ (1993) *Principles and techniques of contemporary taxonomy.* Blackie Academic Publishers, London.
4. Fleure HJ (1954) In *The distribution of the human blood groups* (A Mourant), p. xix. Blackwell Scientific Publications.
5. Zegura SL *et al.* (1990) *Current Anthropology* **32**: 420–426.
6. Cann RL, Stoneking M, Wilson AC (1987) *Nature* **325**: 31–36.
7. Gamble C (1997) Speaking at 'Science and Archaeology' organised by English Heritage. London. February 1997.
8. Jobling MA, Tyler-Smith C (1995) *Trends in Genetics* **11**: 449–456.
9. Sykes BC *et al.* (1995) *American Journal of Human Genetics* **57**:1463–1475.
10. Ammerman AJ, Cavalli-Sforza LL (1984) *The Neolithic transition and the genetics of populations in Europe.* Princeton University Press, Princeton.
11. Renfrew C. (1996) in *The origins and spread of agriculture and pastoralism in Eurasia* (DR Harris ed.), p. 86. UCL Press. London.
12. Richards M *et al.* (1996) *American Journal of Human Genetics* **59**: 185–203.
13. Mourant A (ed.) (1954) 'The distribution of the human blood groups'. Blackwell Scientific Publications, Oxford.
14. Cavalli-Sforza LL, Minch E (1997) *American Journal of Human Genetics* **61**: 247–251.
15. Krings M *et al.* (1997) *Cell* **90**: 19–30.

CHAPTER 6

1. Stone AC and Stoneking M (1993) Ancient DNA from a pre-Colombian Amerindian population. *Am. J. Phys. Anthropol.* **92(4)**: 463–71.
2. Handt O, Krings M, Ward RH and Paabo S. (1996) The retrieval of ancient human DNA sequences. *Am. J. Hum. Genet.* **59(2)**: 368–76.
3. Cooper A *et al.* (1993). Independent origins of New Zealand moas and kiwis. *Proc. Natl. Acad. Sci. USA* **89**: 8741–4.
4. Woodward SR *et al.* (1994) DNA Sequence from Cretaceous period bone fragments. *Science* **266**: 1229–32.
5. Zischler H *et al.* (1995) A nuclear 'fossil' of the mitochondrial D-loop and the origin of modern humans. *Nature* **378**: 489–92.
6. Poinar HN *et al.* (1996) Amino acid racemization and the preservation of ancient DNA. *Science* **272**: 864–6.
7. Golenberg EM *et al.* (1990) Chloroplast DNA from a Miocene Magnolia species. *Nature* **344**: 656–8.
8. Austin JJ *et al.* (1997) Problems of reproducibility—does geologically ancient DNA survive in amber-preserved insects. *Proc. R. Soc. Lond. B. Biol. Sci.* **264**: 467–74.

CHAPTER 7

1. Ward RH *et al.* (1991) Extensive mitochondrial diversity within a single Amerindian tribe. *Proc. Natl. Acad. Sci. USA* **88**; 8720–8724.
2. Kolman *et al.* (1996) Mitochondrial DNA analysis of Mongolian populations and implications for the origin of New World founders. *Genetics* **142**; 1321–1334.
3. Harding *et al.* (1997) Archaic African *and* Asian lineages in the genetic ancestry of modern humans. *Am. J. Hum. Genet.* **60**; 772–789.
4. Ward RH *et al.* (1993) Genetic and linguistic differentiation in the Americas. *Proc. Natl. Acad. Sci. USA* **90**; 10663–10667.
5. Kolman CJ *et al.* (1995) Reduced mtDNA diversity in the Ngobe Amerinds of Panama. *Genetics* **140**; 275–283.

CHAPTER 8

1. Hill HF, Hill AG, Bodmer JG (1976). Clinical diagnosis of ankylosing spondylitis in women and relation to presence of HLA-B27. *Ann. Rheum. Dis* **35**; 267–270.
2. Bjorkman PJ *et al.* (1987) The foreign antigen binding site and T cell recognition regions of class I histocompatibility antigens. *Nature* **329**; 512–518.

Index